T0264701

# Modeling Shallow Water Flows Using the Discontinuous Galerkin Method

# Modeling Shallow Water Flows Using the Discontinuous Galerkin Method

Abdul A. Khan
Wencong Lai

**CRC Press**
Taylor & Francis Group
Boca Raton  London  New York

CRC Press is an imprint of the
Taylor & Francis Group, an **informa** business

CRC Press
Taylor & Francis Group
6000 Broken Sound Parkway NW, Suite 300
Boca Raton, FL 33487-2742

First issued in paperback 2017

© 2014 by Taylor & Francis Group, LLC
CRC Press is an imprint of Taylor & Francis Group, an Informa business

No claim to original U.S. Government works

ISBN-13: 978-1-4822-2601-0 (hbk)
ISBN-13: 978-1-138-07646-4 (pbk)

This book contains information obtained from authentic and highly regarded sources. Reasonable efforts have been made to publish reliable data and information, but the author and publisher cannot assume responsibility for the validity of all materials or the consequences of their use. The authors and publishers have attempted to trace the copyright holders of all material reproduced in this publication and apologize to copyright holders if permission to publish in this form has not been obtained. If any copyright material has not been acknowledged please write and let us know so we may rectify in any future reprint.

Except as permitted under U.S. Copyright Law, no part of this book may be reprinted, reproduced, transmitted, or utilized in any form by any electronic, mechanical, or other means, now known or hereafter invented, including photocopying, microfilming, and recording, or in any information storage or retrieval system, without written permission from the publishers.

For permission to photocopy or use material electronically from this work, please access www.copyright.com (http://www.copyright.com/) or contact the Copyright Clearance Center, Inc. (CCC), 222 Rosewood Drive, Danvers, MA 01923, 978-750-8400. CCC is a not-for-profit organization that provides licenses and registration for a variety of users. For organizations that have been granted a photocopy license by the CCC, a separate system of payment has been arranged.

**Trademark Notice:** Product or corporate names may be trademarks or registered trademarks, and are used only for identification and explanation without intent to infringe.

**Visit the Taylor & Francis Web site at
http://www.taylorandfrancis.com**

**and the CRC Press Web site at
http://www.crcpress.com**

# Contents

# *Preface*

Computational modeling has become an integral part of the planning, design, and assessment of water resources projects in general, and river hydraulics in particular. Most of the river flows and near coastal zones belong to a class of flows known as shallow water flow. The governing equations for such flows are hyperbolic in nature and require special attention in terms of numerical modeling. The numerical schemes applied to such flows must be capable of capturing shocks and be able to handle wet/dry fronts. There is a vast amount of literature available that provides in-depth reviews of modeling shallow water flows using the finite difference, finite volume, and finite element methods. New techniques within these fields are being developed as the field of computational modeling is continuously evolving. At the same time, new computational methods are being developed that hold promise in modeling shallow water flows. One such method is the discontinuous Galerkin method, which is being presented in this book.

This book introduces the discontinuous Galerkin method for modeling shallow flows and provides the necessary background information for implementing the method. One- and two-dimensional shallow water flows are considered and represent depth-averaged flow equations (cross-section averaged in the case of one-dimensional natural channel flows). The different forms in which the shallow water flow equations can be written are presented. Step-by-step implementation details are also provided. The authors envision the book to be used in graduate-level courses, and by engineers and researchers.

One of the important aspects of this book is the collection of tests that the authors have gone to great lengths to collect. These tests show the validity of the discontinuous Galerkin scheme in modeling shallow water flows. Readers can use the tests and results for comparison as they develop the discontinuous Galerkin scheme. In addition, the set of tests provides a good platform for validating new schemes.

# Authors

**Abdul A. Khan, Ph.D.,** is an associate professor in the Glenn Department of Civil Engineering at Clemson University (South Carolina). He received a Ph.D. from the University of Alberta, Edmonton, Canada. After his Ph.D., Dr. Khan worked at the National Center for Computational Hydroscience and Engineering, University of Mississippi, before moving to Clemson University. He has been working in the area of computational modeling of river hydraulics, dam-break flows, and sediment transport for the past 20 years. He has published close to 50 journal articles related to his research work and several papers on river flood and dam-break flow modeling. Dr. Khan recently coedited a book, *Sediment Transport Monitoring, Modeling and Management* (Nova Science Publishers, 2013).

**Wencong Lai, Ph.D.,** earned a Ph.D. (2012) and an M.S. (2010) in civil engineering from Clemson University, in the area of applied fluid mechanics, and a B.E. (2008) in water conservancy and hydropower engineering from Huazhong University of Science and Technology, China. He is currently a postdoctorate research associate at the University of Wyoming and a member of the CI-WATER's High-Resolution Multi-Physics Watershed Modeling team. Dr. Lai's research focuses on computational hydraulics and hydrology. He has developed numerical models for 1D and 2D shallow water flows in natural rivers and watersheds using the discontinuous Galerkin finite element method. Dr. Lai is a member of the American Geophysical Union (AGU).

# *chapter one*

# *Introduction*

Computational modeling is integral to the analysis, design, and assessment of river hydraulics. Finite difference, finite volume, and finite element methods for modeling river flows are common. These methods have advantages and disadvantages. In recent decades, the discontinuous Galerkin (DG) method, or so-called discontinuous finite element method, has received great attention in computational fluid dynamics (CFD). The DG method combines the advantages of finite volume and finite element methods. In this book, the DG method and its application to shallow water flows are introduced. A brief historical overview of the DG method and a summary of the chapters in this book are presented in this chapter.

## 1.1 A historical overview

The DG finite element method was first introduced by Reed and Hill (1973) for solving the steady-state neutron transport equation. The approximate solution was computed element by element. The elements were ordered based on the characteristic direction due to the linear nature of the equation. For a locally smooth approximation of degree $k$, LeSaint and Raviart (1974) proved that the convergence rate was of the order $(\Delta x)^k$ for grids with triangular elements and $(\Delta x)^{k+1}$ for Cartesian grids. Johnson and Pitkäranta (1986) further proved that the method had a convergence rate of $(\Delta x)^{k+1/2}$ on general grids, and Peterson (1991) confirmed that it was the optimal convergence rate. Analysis of linear problems with nonsmooth solutions was performed by Lin and Zhou (1993).

Early applications of the DG method involved wave propagation in elastic media (Wellford and Oden, 1975), modeling parabolic equations (Jamet, 1978), and simulating viscoelastic flows (Fortin and Fortin, 1989). Chavent and Salzano (1982) applied the DG method to one-dimensional nonlinear scalar conservation laws, with piecewise linear elements in the DG space and the forward Euler method for time discretization. Their explicit scheme was unconditionally unstable unless a very restrictive time step was used. To overcome this problem, Chavent and Cockburn (1987) introduced a suitable slope limiter, thus obtaining a total variation diminishing in the means (TVDM) scheme and total variation bounded (TVB) scheme. These schemes were stable for Courant–Friedrichs–Lewy (CFL) numbers less than or equal to 1/2. However, these schemes were only first-order accurate in time and the slope limiter affected the quality of solution in smooth regions. The problem was rectified by the introduction of the Runge–Kutta discontinuous Galerkin (RKDG) method (Cockburn and Shu, 1988), where the second-order total variation diminishing (TVD) Runge–Kutta method and an improved slope limiter (Shu, 1987) were combined. The resulting explicit RKDG method was found to be linearly stable for CFL numbers less than 1/3, maintained formal accuracy in smooth regions, and ensured sharp shock resolution without oscillations and convergence to entropy solutions even for nonconvex fluxes.

Cockburn and Shu (1989) extended this approach to high-order RKDG methods for scalar conservation laws with general slope limiters and high-order, nonlinear stable Runge–Kutta methods. These RKDG methods were further extended to one-dimensional (Cockburn and Lin, 1989) and multidimensional (Cockburn et al., 1990; Cockburn and Shu, 1998a) systems with slope limiters. In multidimensional space problems, the development of high-order TVD schemes is not as straightforward as in one-dimensional cases. Goodman and LeVeque (1985) proved that any TVD scheme was at most first-order accurate. Thus, the TVD slope limiters in higher dimensions would reduce the accuracy of the scheme to first order. This was overcome by introducing a general slope limiter (Cockburn and Shu, 1998a) that enforced a local maximum principle only, which was compatible with high-order accuracy.

The RKDG method is a combination of the finite volume method (FVM) and the finite element method (FEM). As a result, it keeps the advantages of both FVM and FEM. First, the RKDG method can handle complex geometries and boundary conditions. Second, it can provide high-order approximations through the use of high-order interpolating functions. Third, it yields element-wise local formulation for which global matrix assembly is not needed and is highly parallelizable. Also, the *h*- and *p*-adaptive refinements can be made easily due to its local formulation. Last, the RKDG is efficient for convection-dominated flows as various upwinding schemes can be easily incorporated through the discontinuous element boundaries.

However, the DG method has its drawbacks, such as a larger number of variables compared to the continuous FEM, and it is inefficient to handle higher-order spatial terms. In the application of the RKDG method to the nonlinear second-order hyperbolic equations, Chen et al. (1995) approximated the second-order terms by simple projections into suitable finite element space. Bassi and Rebay (1997) applied the original idea of RKDG method to the compressible Navier–Stokes equations, where the second-order derivatives were reduced to first order by using intermediate variables. Cockburn and Shu (1998b) generalized these methods and introduced the local discontinuous Galerkin (LDG) method. The basic idea of the LDG method is to rewrite the original system into a larger, degenerate, first-order system. Thus, the higher-order parabolic and elliptic problems can be handled with the LDG methods. The TVD Runge–Kutta methods of order up to 4 can be found in Gottlieb and Shu (1998). In addition to the Runge–Kutta methods, other methods can be used for time discretization such as the Lax–Wendroff scheme (Titarev and Toro, 2002).

Using the DG method to model shocks, slope limiters are needed to suppress spurious oscillations. Many slope limiters exist in the literature (Cockburn and Shu, 1989; Burbeau et al., 2001; Tu and Aliabadi, 2005; Krivodonova, 2007). In recent years, the essentially nonoscillatory (ENO), weighted ENO (WENO), or Hermite WENO (HWENO) schemes (Qiu and Shu, 2005a,b; Luo et al., 2007) have been used as slope limiters in the DG method. The ENO (Shu and Osher, 1988) and WENO (Liu et al., 1994) methods were developed in finite difference and finite volume frameworks to achieve high-order accuracy and sharp shock transition. In high-order schemes, the ENO/WENO methodology is more convenient and robust than the slope limiter methodology.

Lately, the general $P_N P_M$ approach is proposed to provide a unified framework for the construction of finite volume and DG methods (Dumbser et al., 2008; Dumbser, 2010). A comprehensive theoretical development, numerical applications, and recent research about the DG method can be found in literature (Cockburn et al., 1998; Hesthaven and Warburton, 2008; Wang, 2011; Lai and Khan, 2011a,b; Lai and Khan, 2012a,b; Lai and Khan, 2013).

## 1.2 Organization of the book

As the discontinuous Galerkin method is an efficient tool for first-order hyperbolic conservation laws, the focus of this book will be on such types of problems, especially one- and two-dimensional shallow water flows. The discussion in the following chapters highlights the numerical implementation of the DG method. In Chapter 2, the discontinuous Galerkin procedure for hyperbolic conservation laws is introduced. The key ideas of the DG method are explained. In addition, a short overview of some mathematical preliminaries is presented, such as shape functions, isoparametric mapping, numerical integration, approximate Riemann solvers, and time integration.

To provide a better insight into the DG method, some numerical tests with the DG method for one-dimensional, nonconservative problems are presented in Chapter 3. These tests include the first-order ordinary differential equation (ODE), linear convection, and second-order transient and steady diffusion. The basic formulation and unique qualities of the DG method are shown through these numerical examples.

Starting from Chapter 4, the focus is on the DG method for the conservation laws, especially shallow water flows. In Chapter 4, the DG method is applied to one-dimensional scalar equations and the system of hyperbolic conservation laws, that is, the Burgers' equation and shallow water flow equations. More features of the DG method are investigated, such as the TVD slope limiter and numerical flux. In Chapter 5, the DG method is applied to the one-dimensional shallow water flows in nonrectangular, nonprismatic channels. Discussion about the choice of shallow water flow equations and its effect are also presented. In Chapters 6 through 8, the DG scheme is applied to the two-dimensional shallow water flows. Last, numerical simulations for shallow water flows with pollutant transport in one and two dimensions are presented in Chapter 9. Concluding remarks are presented in Chapter 10.

## *References*

Bassi, F., and Rebay, S. (1997). A high-order accurate discontinuous finite element method for the numerical solution of the compressible Navier-Stokes equations. *Journal of Computational Physics*, 131(2), 267–279.

Burbeau, A., Sagaut, P., and Bruneau, C.-H. (2001). A problem-independent limiter for high-order Runge–Kutta discontinuous Galerkin methods. *Journal of Computational Physics*, 169(1), 111–150.

Chavent, G., and Cockburn, B. (1987). The local projection $P^0P^1$-Discontinuous-Galerkin finite element method for scalar conservation laws. *IMA Preprint Series 341*, University of Minnesota.

Chavent, G., and Salzano, G. (1982). A finite-element method for the 1D water flooding problem with gravity. *Journal of Computational Physics*, 45(3), 307–344.

Chen, Z., Cockburn, B., Jerome, J., and Shu, C. W. (1995). Mixed-RKDG finite element methods for the 2-D hydrodynamic model for semiconductor device simulation. *VLSI Design*, 3(2), 145–158.

Cockburn, B., Hou, S., and Shu, C. W. (1990). The Runge–Kutta local projection discontinuous Galerkin finite element method for conservation laws IV: The multidimensional case. *Mathematics of Computation*, 54(190), 545–581.

Cockburn, B., Karniadakis, G. E., and Shu, C. W. (1998). *Discontinuous Galerkin Methods: Theory, Computation and Application*. Springer, Berlin.

Cockburn, B., and Lin, S. Y. (1989). TVB Runge–Kutta local projection discontinuous Galerkin finite element method for conservation laws III: One dimensional systems. *Journal of Computational Physics*, 84(1), 90–113.

Cockburn, B., and Shu, C. W. (1988). The Runge–Kutta local projection $P^1$-discontinuous-Galerkin finite element method for scalar conservation laws. *IMA Preprint Series 388*, University of Minnesota.

Cockburn, B., and Shu, C. W. (1989). TVB Runge–Kutta local projection discontinuous Galerkin finite element method for conservation laws II: General framework. *Mathematics of Computation*, 52(186), 411–435.

Cockburn, B., and Shu, C. W. (1998a). The Runge–Kutta discontinuous Galerkin method for conservation laws V: Multidimensional systems. *Journal of Computational Physics*, 141(2), 199–224.

Cockburn, B., and Shu, C. W. (1998b). The local discontinuous Galerkin method for time-dependent convection-diffusion systems. *SIAM Journal on Numerical Analysis*, 35(6), 2440–2463.

Dumbser, M. (2010). Arbitrary high order $P_NP_M$ schemes on unstructured meshes for the compressible Navier-Stokes equations. *Computer & Fluids*, 39(1), 60–76.

Dumbser, M., Balsara, D. S., Toro, E. F., and Munz, C.-D. (2008). A unified framework for the construction of one-step finite volume and discontinuous Galerkin schemes on unstructured meshes. *Journal of Computational Physics*, 227(18), 8209–8253.

Fortin, M., and Fortin, A. (1989). A new approach for the FEM simulation of viscoelastic flows. *Journal of Non-Newtonian Fluid Mechanics*, 32(3), 295–310.

Goodman, J. B., and LeVeque, R. J. (1985). On the accuracy of stable schemes for 2D scalar conservation laws. *Mathematics of Computation*, 45(171), 15–21.

Gottlieb, S., and Shu, C. W. (1998). Total variation diminishing Runge–Kutta schemes. *Mathematics of Computation*, 67(221), 73–85.

Hesthaven, J. S., and Warburton, T. (2008). *Nodal Discontinuous Galerkin Methods: Algorithms, Analysis, and Application*. Springer, New York.

Jamet, P. (1978). Galerkin-type approximations which are discontinuous in time for parabolic equations in a variable domain. *SIAM Journal on Numerical Analysis*, 15(5), 912–928.

Johnson, C., and Pitkäranta, J. (1986). An analysis of the discontinuous Galerkin method for a scalar hyperbolic equation. *Mathematics of Computation*, 46(173), 1–26.

Krivodonova, L. (2007). Limiters for high-order discontinuous Galerkin methods. *Journal of Computational Physics*, 226(1), 879–896.

Lai, W., and Khan, A. A. (2011a). Discontinuous Galerkin method for 1D shallow water flow with water surface slope limiter. *International Journal of Civil and Environmental Engineering*, 3(3), 167–176.

Lai, W., and Khan, A. A. (2011b). A discontinuous Galerkin method for two-dimensional shock wave modeling. *Modelling and Simulation in Engineering*, DOI:10.1155/2011/782832.

Lai, W., and Khan, A. A. (2012a). Discontinuous Galerking method for 1D shallow water flows in natural rivers. *Engineering Applications of Computational Fluid Mechanics*, 6(1), 74–86.

Lai, W., and Khan, A. A. (2012b). Modeling dam-break flood in natural rivers with discontinuous Galerkin method. *Journal of Hydrodynamics*, 24(4), 467–478.

Lai, W., and Khan, A. A. (2013). Time stepping in discontinuous Galerkin method. *Journal of Hydrodynamics*, 25(3), 321–329.

LeSaint, P., and Raviart, P. A. (1974). On a finite element method for solving the neutron transport equation. In *Mathematical Aspects of Finite Elements in Partial Differential Equations*, edited by C. Boor, 89–145, Academic Press, New York.

Lin, Q., and Zhou, A. H. (1993). Convergence of the discontinuous Galerkin methods for a scalar hyperbolic equation. *Acta Mathematica Scientia*, 13, 207–210.

Liu, X., Osher, S., and Chan, T. (1994). Weighted essentially non-oscillatory schemes. *Journal of Computational Physics*, 115, 200–212.

Luo, H., Baum, J. D., and Löhner, R. (2007). A Hermite WENO-based limiter for discontinuous Galerkin method on unstructured grids. *Journal of Computational Physics*, 225(1), 686–713.

Peterson, T. E. (1991). A note on the convergence of the discontinuous Galerkin method for a scalar hyperbolic equation. *SIAM Journal on Numerical Analysis*, 28(1), 133–140.

Qiu, J., and Shu, C. W. (2005a). Runge–Kutta discontinuous Galerkin method using WENO limiters. *SIAM Journal on Scientific Computing*, 26(3), 907–929.

Qiu, J., and Shu, C. W. (2005b). Hermite WENO schemes and their application as limiters for Runge–Kutta discontinuous Galerkin method II: Two dimensional case. *Computers & Fluids*, 34(6), 642–663.

Reed, W. H., and Hill, T. R. (1973). Triangular mesh method for the neutron transport equation. Los Alamos Scientific Laboratory Report, LA-UR-73-479, 1–23.

Shu, C. W. (1987). TVB uniformly high-order schemes for conservation laws.
*Mathematics of Computation*, 49(179), 105–121.

Shu, C. W., and Osher, S. (1998). Efficient implementation of essentially non-
oscillatory shock-capturing schemes. *Journal of Computational Physics*, 77(2),
439–471.

Titarev, V. A., and Toro, E. F. (2002). ADER: Arbitrary high order Godunov
approach. *Journal of Scientific Computing*, 17(1–4), 609–618.

Tu, S., and Aliabadi, S. (2005). A slope limiting procedure in discontinuous Galerkin
finite element method for gasdynamics applications. *International Journal of
Numerical Analysis and Modeling*, 2(2), 163–178.

Wang, Z. J. (2011). *Adaptive High-Order Methods in Computational Fluid Dynamics:
Advances in Computational Fluid Dynamics*. Vol. 2. World Scientific Publishing,
Singapore.

Wellford, L. C., and Oden, J. T. (1975). Discontinuous finite element approximations for the analysis of shock waves in nonlinear elastic materials. *Journal of
Computational Physics*, 19(2), 179–210.

*chapter two*

# General formulation of the discontinuous Galerkin method

In this chapter, the general procedure of the discontinuous Galerkin (DG) method for hyperbolic conservation laws is presented. Before delving in the actual method, a brief review of some mathematical preliminaries relevant to the DG method is provided. Different methods for the treatment of the flux term are outlined. Time integration procedures relevant to the DG formulation are presented.

## 2.1   Conservation form of equations

A system of $m$ partial differential equations in the conservation form can be written as Equation (2.1), where the variables are defined in Equation (2.2). In these equations, $\mathbf{U}$ is the vector of conserved variables, $\mathbf{F}$ is the flux vector, and $\mathbf{S}$ is the vector of source terms. Any flux component, $\mathbf{f}_i$, can be written in a three-dimensional Cartesian coordinate system as shown in Equation (2.3).

$$\frac{\partial \mathbf{U}}{\partial t} + \nabla \cdot \mathbf{F}(\mathbf{U}) = \mathbf{S}(\mathbf{U}), \quad \mathbf{U}(\mathbf{x}, 0) = \mathbf{U}_o(\mathbf{x}), \quad \mathbf{x} \in \Omega, \quad t \geq 0 \tag{2.1}$$

$$\mathbf{U} = \begin{bmatrix} U_1 \\ U_2 \\ \vdots \\ U_m \end{bmatrix}, \quad \mathbf{F}(\mathbf{U}) = \begin{bmatrix} \mathbf{f}_1 \\ \mathbf{f}_2 \\ \vdots \\ \mathbf{f}_m \end{bmatrix}, \quad \mathbf{S}(\mathbf{U}) = \begin{bmatrix} S_1 \\ S_2 \\ \vdots \\ S_m \end{bmatrix} \tag{2.2}$$

$$\mathbf{f}_i = E_i \mathbf{i} + G_i \mathbf{j} + H_i \mathbf{k} \tag{2.3}$$

For an arbitrary unit vector $\mathbf{n} = (n_x, n_y, n_z)$, the Jacobian matrix of the flux function $\mathbf{F}(\mathbf{U})$ is given by Equation (2.4). The system is hyperbolic if the Jacobian matrix has $m$ real eigenvalues, $\lambda_i(\mathbf{U})$, $i = 1, 2, \ldots, m$, and a complete set of linearly independent eigenvectors $\mathbf{K}_i(\mathbf{U})$, $i = 1, 2, \ldots, m$. The system is strictly hyperbolic if the eigenvalues are all real and distinct (Toro, 2009).

$$J(\mathbf{U}) = \frac{\partial \mathbf{F}(\mathbf{U}) \cdot \mathbf{n}}{\partial \mathbf{U}} = \begin{bmatrix} \partial \mathbf{f}_1/\partial U_1 & \cdots & \partial \mathbf{f}_1/\partial U_m \\ \partial \mathbf{f}_2/\partial U_1 & \cdots & \partial \mathbf{f}_2/\partial U_m \\ \vdots & \vdots & \vdots \\ \partial \mathbf{f}_m/\partial U_1 & \cdots & \partial \mathbf{f}_m/\partial U_m \end{bmatrix} \cdot \mathbf{n} \qquad (2.4)$$

### 2.1.1 Discontinuous Galerkin formulation

As customary in finite element formulations, the problem domain $\Omega$ is divided into a collection of *Ne* elements as shown in Equation (2.5). The differences between continuous and discontinuous linear elements in 1D and 2D are shown in Figures 2.1 and 2.2, respectively. In the continuous finite element method, elements share nodes and boundaries with adjoining elements, and discrete variables at common nodes have the same value in all the adjoining elements. While in the discontinuous Galerkin

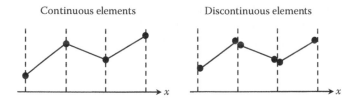

*Figure 2.1* Continuous and discontinuous linear elements in 1D.

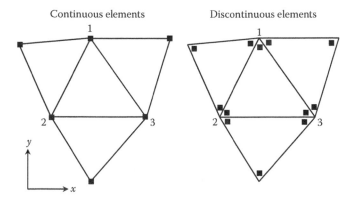

*Figure 2.2* Continuous and discontinuous linear elements in 2D.

method, each element has its own vertices with discrete variables and discontinuities may form at element boundaries.

$$\Omega \approx \hat{\Omega} = \bigcup_{e=1}^{Ne} \Omega_e \tag{2.5}$$

Inside an element, the variation of the conserved variables, fluxes, and source terms can be approximated using the Lagrange polynomials (also called shape functions or basis functions) as given by Equation (2.6). The conservation laws are multiplied with a test function and the resulting equations are integrated over an element as shown in Equation (2.7). In the Galerkin method, the test functions are the same as the shape functions.

$$\left. \begin{aligned} \mathbf{U} \approx \hat{\mathbf{U}} = \sum \mathbf{N}_j(\mathbf{x}) \mathbf{U}_j(\mathbf{x}, t) \\ \mathbf{F}(\mathbf{U}) \approx \mathbf{F}(\hat{\mathbf{U}}) \\ \mathbf{S}(\mathbf{U}) \approx \mathbf{S}(\hat{\mathbf{U}}) \end{aligned} \right\} \tag{2.6}$$

$$\int_{\Omega_e} \mathbf{N}_i \frac{\partial \hat{\mathbf{U}}}{\partial t} d\Omega + \int_{\Omega_e} \mathbf{N}_i \nabla \cdot \mathbf{F}(\hat{\mathbf{U}}) d\Omega = \int_{\Omega_e} \mathbf{N}_i \mathbf{S}(\hat{\mathbf{U}}) d\Omega \tag{2.7}$$

Substituting the approximation of $\mathbf{U}$ and applying the divergence theorem results in Equation (2.8). The flux term $(\mathbf{F}(\mathbf{U}) \cdot \mathbf{n})$ is replaced by the numerical flux $(\tilde{\mathbf{F}})$, and the resulting equation is given by Equation (2.9). After integration, for example, by the Gaussian quadrature rule, the final equation can be written as Equation (2.10) or Equation (2.11), where the mass matrix $(\mathbf{M})$ is given by Equation (2.12).

$$\int_{\Omega_e} \mathbf{N}_i \mathbf{N}_j \, d\Omega \frac{\partial \mathbf{U}_j}{\partial t} + \int_{\Gamma_e} \mathbf{N}_i \mathbf{F}(\hat{\mathbf{U}}) \cdot \mathbf{n} \, d\Gamma - \int_{\Omega_e} \nabla \mathbf{N}_i \cdot \mathbf{F}(\hat{\mathbf{U}}) d\Omega = \\ \int_{\Omega_e} \mathbf{N}_i \mathbf{S}(\hat{\mathbf{U}}) d\Omega \tag{2.8}$$

$$\int_{\Omega_e} \mathbf{N}_i \mathbf{N}_j \, d\Omega \frac{\partial \mathbf{U}_j}{\partial t} + \int_{\Gamma_e} \mathbf{N}_i \tilde{\mathbf{F}} \, d\Gamma - \int_{\Omega_e} \nabla \mathbf{N}_i \cdot \mathbf{F}(\hat{\mathbf{U}}) d\Omega = \int_{\Omega_e} \mathbf{N}_i \mathbf{S}(\hat{\mathbf{U}}) d\Omega \tag{2.9}$$

$$\mathbf{M} \frac{\partial \mathbf{U}}{\partial t} = \mathbf{R} \tag{2.10}$$

$$\frac{\partial \mathbf{U}}{\partial t} = \mathbf{M}^{-1} \mathbf{R} = \mathbf{L} \tag{2.11}$$

$$\mathbf{M} = \int_{\Omega_e} \mathbf{N}_i \mathbf{N}_j \, d\Omega \tag{2.12}$$

Finally, the solution of conserved variables **U** can be obtained using an appropriate time integration method. The continuous conservation laws are broken into discrete algebraic equations and solved to obtain solutions at different times. In a hyperbolic system, discontinuities and shock waves may form even with smooth initial and boundary conditions. Numerical tests show that spurious oscillations are generated with higher-order space approximation, so limiting techniques on conserved variables need to be applied to limit the solution. The form of shape functions, numerical integration, numerical flux, and time integration will be discussed in the following sections.

### 2.1.2   Numerical flux

Since the discontinuous elements are connected through the numerical flux along the boundaries, the accuracy with which these fluxes are calculated becomes crucial in the DG method. As the normal flux $\mathbf{F}(\mathbf{U}) \cdot \mathbf{n}$ is not defined on the discontinuous boundaries, the usual strategy is to replace it with a numerical flux $\tilde{\mathbf{F}}$. Rewriting Equation (2.8), in the form given by Equation (2.13), and integrating it from time $t$ to $t + \Delta t$ results in Equation (2.14).

$$\int_{\Omega_e} \mathbf{N}_i \mathbf{N}_j \, d\Omega \frac{\partial \mathbf{U}_j}{\partial t} + \int_{\Gamma_e} \mathbf{N}_i \mathbf{F}(\hat{\mathbf{U}}) \cdot \mathbf{n} \, d\Gamma = \mathbf{R}' \tag{2.13}$$

$$\int_{\Omega_e} \mathbf{N}_i \mathbf{N}_j \, d\Omega \int_t^{t+\Delta t} \frac{\partial \mathbf{U}_j}{\partial t} dt + \int_{\Gamma_e} \mathbf{N}_i \int_t^{t+\Delta t} \mathbf{F}(\hat{\mathbf{U}}) \cdot \mathbf{n} \, dt \, d\Gamma = \int_t^{t+\Delta t} \mathbf{R}' dt \tag{2.14}$$

The normal flux can be replaced by the numerical flux as shown in Equation (2.15), where the numerical flux is given by Equation (2.16). The numerical flux is the time average of normal flux. The accuracy with which the numerical flux is calculated restricts the time step size. As the normal flux includes quantities such as mass, momentum, and energy flow in or out of an element, the accurate representation of numerical flux is essential to conserve these quantities.

$$\int_t^{t+\Delta t} \mathbf{F}(\hat{\mathbf{U}}) \cdot \mathbf{n} \, dt = \int_t^{t+\Delta t} \tilde{\mathbf{F}} \, dt = \tilde{\mathbf{F}} \Delta t \tag{2.15}$$

$$\tilde{\mathbf{F}} = \frac{1}{\Delta t} \int_t^{t+\Delta t} \mathbf{F}(\hat{\mathbf{U}}) \cdot \mathbf{n} \, dt \tag{2.16}$$

The solution of the numerical flux depends on the variables **U** in an element $\Omega_e$ and the neighboring element $\Omega_{nb}$, which becomes a general approximate Riemann problem and can be written as shown in Equation (2.17). To conserve mass in the whole domain, the numerical flux is required to be consistent with the physical flux (Laney, 1998) as shown in

Equation (2.18). More discussions regarding numerical flux are provided in Section 2.5.

$$\tilde{\mathbf{F}} = \tilde{\mathbf{F}}(\mathbf{U}_e, \mathbf{U}_{nb}) \tag{2.17}$$

$$\tilde{\mathbf{F}} = \tilde{\mathbf{F}}(\mathbf{U}, \mathbf{U}) = \mathbf{F}(\mathbf{U}) \cdot \mathbf{n} \quad \text{for} \quad \mathbf{U} = \mathbf{U}_e = \mathbf{U}_{nb} \tag{2.18}$$

## 2.2 Shape functions

The shape functions developed from the Lagrangian interpolation theory are briefly discussed in this section. The shape functions (alternatively known as basis functions, interpolation functions, or trial functions) are a piecewise polynomial approximation of the solution in a local element. More discussion about the shape functions can be found in the textbooks dealing with the finite element method (Reddy, 1993; Lewis et al., 2004; Zienkiewicz et al., 2005; Li, 2006).

### 2.2.1 1D Shape functions

The shape functions are piecewise continuous and approximate the variation of variables within an element. In the Lagrangian interpolation, the variation of a variable is approximated using Equation (2.19), where $n$ is the number of nodes in an element, $U_j$ is the solution at the nodes, $x_s^e$ and $x_e^e$ are the start and end coordinates of an element, respectively, and $N_j(x)$ is the $j$th Lagrange basis function given by Equation (2.20).

$$U(x) \simeq \hat{U}(x) = \sum_{j=1}^{n} N_j(x) U_j, \quad x \in \Omega_e = \left[ x_s^e, x_e^e \right] \tag{2.19}$$

$$N_j(x) = \frac{\Pi_{k=1, k \neq j}^{n}(x - x_k)}{\Pi_{k=1, k \neq j}^{n}(x_j - x_k)}, \quad j = 1, 2, \ldots n \tag{2.20}$$

The shape function $N_j(x)$ has a value of one at node $j$ under consideration and zero at all other nodes as given by Equation (2.21). The approximation $\hat{U}$ yields the value of $U$ at each node of the element. For a 2-node linear element, the shape functions are given by Equation (2.22) and are shown in Figure 2.3. The linear variation of a variable is shown in

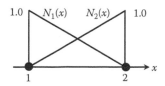

*Figure 2.3* 1D linear shape functions.

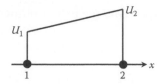

**Figure 2.4** 1D linear variation of a variable in an element.

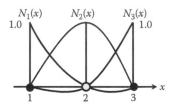

**Figure 2.5** 1D quadratic shape functions.

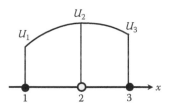

**Figure 2.6** 1D quadratic variation of a variable in an element.

Figure 2.4. For a 3-node quadratic element, the shape functions are given by Equation (2.23) and are shown in Figure 2.5, where $x_i$ ($i$ = 1, 2, 3) is the coordinate of a node. The corresponding quadratic variation of a variable is shown in Figure 2.6.

$$N_j(x_i) = \begin{cases} 1 & i = j \\ 0 & i \neq j \end{cases} \tag{2.21}$$

$$\left. \begin{aligned} N_1(x) &= \frac{x - x_2}{x_1 - x_2} \\ N_2(x) &= \frac{x - x_1}{x_2 - x_1} \end{aligned} \right\} \tag{2.22}$$

$$N_1(x) = \frac{(x - x_2)(x - x_3)}{(x_1 - x_2)(x_1 - x_3)}$$

$$N_2(x) = \frac{(x - x_1)(x - x_3)}{(x_2 - x_1)(x_2 - x_3)} \qquad (2.23)$$

$$N_3(x) = \frac{(x - x_1)(x - x_2)}{(x_3 - x_1)(x_3 - x_2)}$$

### 2.2.2  2D Shape functions

In two-dimensional analysis, the triangular and quadrilateral elements are commonly used. Generally, these elements can have straight or curved edges. Since the triangular elements are more adaptive to complex geometry and require the least number of nodes to achieve a given order of polynomial, the discussion is restricted to triangular elements. The 2D linear triangular elements with straight edges are used throughout this book. The variation of variables can be approximated in an element as given by Equation (2.24), where $n$ is the number of nodes in an element. The shape functions are required to satisfy the Kronecker delta property given by Equation (2.25). The shape function $N_1$ and the variation of a variable over a 3-node linear triangular element with straight edges are shown in Figures 2.7 and 2.8, respectively. The nodes are numbered counterclockwise. The shape functions for linear triangular elements are given by Equation (2.26), where $A$ is the area of the triangular element given by Equation (2.27) and $x_i, y_i$ are the coordinates of a node ($i = 1, \ldots, n$).

$$U(x,y) \simeq \hat{U}(x,y) = \sum_{j=1}^{n} N_j(x,y) U_j, \quad (x,y) \in \Omega_e \qquad (2.24)$$

$$N_j(x_i, y_i) = \begin{cases} 1 & \text{if} \quad i = j \\ 0 & \text{if} \quad i \neq j \end{cases} \qquad (2.25)$$

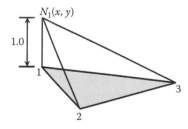

*Figure 2.7* Shape function for a linear triangular element.

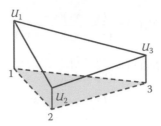

**Figure 2.8** Linear variation of a variable in a triangular element.

$$\left. \begin{aligned} N_1(x,y) &= \frac{(x_2 y_3 - x_3 y_2) + (y_2 - y_3)x + (x_3 - x_2)y}{2A} \\ N_2(x,y) &= \frac{(x_3 y_1 - x_1 y_3) + (y_3 - y_1)x + (x_1 - x_3)y}{2A} \\ N_3(x,y) &= \frac{(x_1 y_2 - x_2 y_1) + (y_1 - y_2)x + (x_2 - x_1)y}{2A} \end{aligned} \right\} \tag{2.26}$$

$$A = \frac{1}{2} \begin{vmatrix} 1 & x_1 & y_1 \\ 1 & x_2 & y_2 \\ 1 & x_3 & y_3 \end{vmatrix} \tag{2.27}$$

## 2.3　Isoparametric mapping

The calculation of mass matrix in Equation (2.10) involves integration, which can be calculated analytically for 1D problems with lower-order inter-polation and test functions. However, in 2D and 3D problems, especially with higher-order elements and curved edges, the integrals can no longer be evaluated in closed form. The mass matrix calculations are facilitated through the concept of isoparametric mapping and numerical quadrature. The isoparametric mapping is also called transformed coordinates, and these transformed coordinates enable the use of numerical quadrature to evaluate integrals.

In isoparametric mapping, both the variation of variables and element geometry are represented by the same shape functions. The shape functions are based on the normalized coordinate system. The element defined in the normalized coordinate system is referred to as the canonical element or master element. The mapping from the

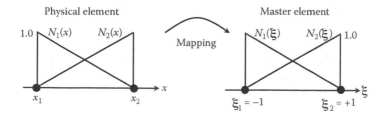

**Figure 2.9** Isoparametric mapping for a 1D linear element.

canonical element to the physical element takes the form shown by Equation (2.28).

$$x = \sum N_j(\xi)x_j, \quad \hat{U} = \sum N_j(\xi)U_j \qquad (2.28)$$

In this section, the isoparametric mappings in 1D and 2D are explained. The mapping from a physical element to the canonical element is also illustrated, which is referred to as the transformation from a global to local coordinates system. The details of integration by numerical quadrature will be provided in the next section.

## 2.3.1   Isoparametric mapping in 1D

The 1D isoparametric mapping in general form is given by Equation (2.29). For a linear element, the isoparametric mapping is shown in Figure 2.9 and the shape functions in local coordinate are given by Equation (2.30). The variation of the global coordinate and the variation of any variable within a linear element are given by Equation (2.31).

$$\left. \begin{array}{l} x = \displaystyle\sum_{j=1}^{n} N_j(\xi)x_j \\[4mm] \hat{U} = \displaystyle\sum_{j=1}^{n} N_j(\xi)U_j \end{array} \right\} \qquad (2.29)$$

$$\left. \begin{array}{l} N_1(\xi) = 0.5(1-\xi) \\[2mm] N_2(\xi) = 0.5(1+\xi) \end{array} \right\} \qquad (2.30)$$

$$\left. \begin{array}{l} x = N_1(\xi)x_1 + N_2(\xi)x_2 \\[2mm] \hat{U} = N_1(\xi)U_1 + N_2(\xi)U_2 \end{array} \right\} \qquad (2.31)$$

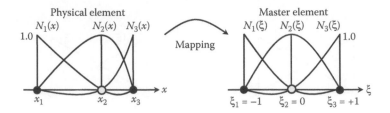

*Figure 2.10* Isoparametric mapping for a 1D quadratic element.

The isoparametric mapping for a 1D quadratic element is shown in Figure 2.10. The shape functions for the quadratic element are given by Equation (2.32). The mapping between the local and global coordinate systems and the variation of any variable within a quadratic element are given by Equation (2.33).

$$
\left.\begin{aligned}
N_1(\xi) &= 0.5\xi(\xi - 1) \\
N_2(\xi) &= (1 + \xi)(1 - \xi) \\
N_3(\xi) &= 0.5\xi(1 + \xi)
\end{aligned}\right\} \tag{2.32}
$$

$$
\left.\begin{aligned}
x &= N_1(\xi)x_1 + N_2(\xi)x_2 + N_3(\xi)x_3 \\
\hat{U} &= N_1(\xi)U_1 + N_2(\xi)U_2 + N_3(\xi)U_3
\end{aligned}\right\} \tag{2.33}
$$

### 2.3.2   Isoparametric mapping in 2D

The isoparametric mapping for a 2D linear triangular element is shown in Figure 2.11. The shape functions for the linear element are given by

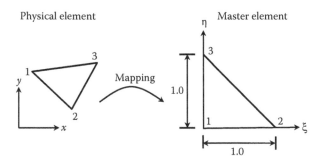

*Figure 2.11* Isoparametric mapping for a 2D linear triangular element.

Equation (2.34). The variable and coordinates can be interpolated within an element by using Equation (2.35).

$$
\left.\begin{array}{l}
N_1(\xi, \eta) = 1 - \xi - \eta \\[6pt]
N_2(\xi, \eta) = \xi \\[6pt]
N_3(\xi, \eta) = \eta
\end{array}\right\} \tag{2.34}
$$

$$
\left.\begin{array}{l}
x = \displaystyle\sum_{j=1}^{3} N_j(\xi, \eta) x_j \\[16pt]
y = \displaystyle\sum_{j=1}^{3} N_j(\xi, \eta) y_j \\[16pt]
\hat{U} = \displaystyle\sum_{j=1}^{3} N_j(\xi, \eta) U_j
\end{array}\right\} \tag{2.35}
$$

### 2.3.3  Integral calculation with isoparametric mapping

Using the isoparametric mapping, the integration can be performed by the method of variable substitution, also called coordinate transformation. The mass matrix in 1D is given by Equation (2.36). The corresponding mass matrix in the local coordinate is given by Equation (2.37). The Jacobian in an element is defined by Equation (2.38). A derivative term in the global coordinate, shown in Equation (2.39), can be transformed into the local coordinate as given by Equation (2.40). Thus, Equation (2.39) in the local coordinate system can be written as Equation (2.41).

$$
\mathbf{M} = \int_{x_s^e}^{x_e^e} N_i(x) N_j(x)\, dx \tag{2.36}
$$

$$
\mathbf{M} = \int_{-1}^{1} N_i(\xi) N_j(\xi) J^e \, d\xi \tag{2.37}
$$

$$
J^e = \frac{dx}{d\xi} = \frac{d\left(\sum_{j=1}^{n} N_j(\xi) x_j\right)}{d\xi} = \sum_{j=1}^{n} \frac{dN_j(\xi)}{d\xi} x_j \tag{2.38}
$$

$$
\int_{x_s^e}^{x_e^e} \frac{dN_i(x)}{dx} f(\hat{U}(x))\, dx \tag{2.39}
$$

$$\frac{dN_i(x)}{dx} \rightarrow \frac{dN_i(\xi)}{d\xi}\frac{d\xi}{dx} = \frac{dN_i(\xi)}{d\xi}\frac{1}{J^e} \tag{2.40}$$

$$\int_{x_s^e}^{x_e^e} \frac{dN_i(x)}{dx} f(\hat{U}(x))dx = \int_{-1}^{1} \frac{dN_i(\xi)}{d\xi}\frac{1}{J^e} f(\hat{U}(\xi))J^e \, d\xi$$

$$= \int_{-1}^{1} \frac{dN_i(\xi)}{d\xi} f(\hat{U}(\xi))d\xi \tag{2.41}$$

In 2D elements, the relationship between the local and global coordinate systems for an infinitesimal area is given by Equation (2.42), where the Jacobian matrix is given by Equation (2.43). To achieve coordinate transformation, the chain rule as shown in Equation (2.44) is used. A derivative in global coordinates can be transformed into a local coordinates system using Equation (2.45).

$$d\Omega = \det(\mathbf{J}^e)d\xi d\eta \tag{2.42}$$

$$\mathbf{J}^e = \begin{bmatrix} \dfrac{\partial x}{\partial \xi} & \dfrac{\partial y}{\partial \xi} \\[2mm] \dfrac{\partial x}{\partial \eta} & \dfrac{\partial y}{\partial \eta} \end{bmatrix} = \begin{bmatrix} \displaystyle\sum_{j=1}^{n} \dfrac{\partial N_j(\xi,\eta)}{\partial \xi} x_j & \displaystyle\sum_{j=1}^{n} \dfrac{\partial N_j(\xi,\eta)}{\partial \xi} y_j \\[4mm] \displaystyle\sum_{j=1}^{n} \dfrac{\partial N_j(\xi,\eta)}{\partial \eta} x_j & \displaystyle\sum_{j=1}^{n} \dfrac{\partial N_j(\xi,\eta)}{\partial \eta} y_j \end{bmatrix} \tag{2.43}$$

$$\begin{bmatrix} \dfrac{\partial}{\partial \xi} \\[2mm] \dfrac{\partial}{\partial \eta} \end{bmatrix} = \begin{bmatrix} \dfrac{\partial x}{\partial \xi} & \dfrac{\partial y}{\partial \xi} \\[2mm] \dfrac{\partial x}{\partial \eta} & \dfrac{\partial y}{\partial \eta} \end{bmatrix} \begin{bmatrix} \dfrac{\partial}{\partial x} \\[2mm] \dfrac{\partial}{\partial y} \end{bmatrix} = \mathbf{J}^e \begin{bmatrix} \dfrac{\partial}{\partial x} \\[2mm] \dfrac{\partial}{\partial y} \end{bmatrix} \tag{2.44}$$

$$\begin{bmatrix} \dfrac{\partial}{\partial x} \\[2mm] \dfrac{\partial}{\partial y} \end{bmatrix} = (\mathbf{J}^e)^{-1} \begin{bmatrix} \dfrac{\partial}{\partial \xi} \\[2mm] \dfrac{\partial}{\partial \eta} \end{bmatrix}, \quad (\mathbf{J}^e)^{-1} = \frac{1}{\det(\mathbf{J}^e)} \begin{bmatrix} \dfrac{\partial y}{\partial \eta} & -\dfrac{\partial y}{\partial \xi} \\[2mm] -\dfrac{\partial x}{\partial \eta} & \dfrac{\partial x}{\partial \xi} \end{bmatrix} \tag{2.45}$$

The procedure of transforming coordinates will be outlined using Equation (2.46), which was derived before as Equation (2.9). The mass matrix can be evaluated using Equation (2.47). The flux term is transformed as shown in Equation (2.48), where the new operator $\nabla'$ is defined by Equation (2.49). Using the isoparametric mapping shown in Equation (2.50), Equation (2.48) can be written in the form given by Equation (2.51), where $\mathbf{F}(\hat{\mathbf{U}}) = \mathbf{E}(\hat{\mathbf{U}})\mathbf{i} + \mathbf{G}(\hat{\mathbf{U}})\mathbf{j}$.

$$\int_{\Omega_e} \mathbf{N}_i \mathbf{N}_j \, d\Omega \frac{\partial \mathbf{U}_j}{\partial t} + \int_{\Gamma_e} \mathbf{N}_i \tilde{\mathbf{F}} \, d\Gamma - \int_{\Omega_e} \nabla \mathbf{N}_i \cdot \mathbf{F}(\hat{\mathbf{U}}) \, d\Omega = \int_{\Omega_e} \mathbf{N}_i \mathbf{S}(\hat{\mathbf{U}}) \, d\Omega \quad (2.46)$$

$$\int_{\Omega_e} \mathbf{N}_i \mathbf{N}_j \, d\Omega = \int_0^1 \int_0^1 \mathbf{N}_i(\xi, \eta) \mathbf{N}_j(\xi, \eta) \det(\mathbf{J}^e) \, d\xi \, d\eta \quad (2.47)$$

$$\int_{\Omega_e} \nabla \mathbf{N}_i \cdot \mathbf{F}(\hat{\mathbf{U}}) \, d\Omega = \int_0^1 \int_0^1 \nabla' \mathbf{N}_i \cdot \mathbf{F}(\hat{\mathbf{U}}) \det(\mathbf{J}^e) \, d\xi \, d\eta \quad (2.48)$$

$$\nabla' = \frac{1}{\det(\mathbf{J}^e)} \left[ \left( \frac{\partial y}{\partial \eta} \frac{\partial}{\partial \xi} - \frac{\partial y}{\partial \xi} \frac{\partial}{\partial \eta} \right) \mathbf{i} + \left( -\frac{\partial x}{\partial \eta} \frac{\partial}{\partial \xi} + \frac{\partial x}{\partial \xi} \frac{\partial}{\partial \eta} \right) \mathbf{j} \right] \quad (2.49)$$

$$\left. \begin{aligned} \frac{\partial y}{\partial \eta} &= y_3 - y_1 = y_{31} \\[2mm] \frac{\partial y}{\partial \xi} &= y_2 - y_1 = y_{21} \\[2mm] \frac{\partial x}{\partial \eta} &= x_3 - x_1 = x_{31} \\[2mm] \frac{\partial x}{\partial \xi} &= x_2 - x_1 = x_{21} \end{aligned} \right\} \quad (2.50)$$

$$\int_{\Omega_e} \nabla \mathbf{N}_i \cdot \mathbf{F}(\hat{\mathbf{U}}) \, d\Omega = \int_0^1 \int_0^1 \left[ y_{31} \frac{\partial \mathbf{N}_i}{\partial \xi} - y_{21} \frac{\partial \mathbf{N}_i}{\partial \eta} \right] \mathbf{E}(\hat{\mathbf{U}}) \, d\xi \, d\eta +$$

$$\int_0^1 \int_0^1 \left[ -x_{31} \frac{\partial \mathbf{N}_i}{\partial \xi} + x_{21} \frac{\partial \mathbf{N}_i}{\partial \eta} \right] \mathbf{G}(\hat{\mathbf{U}}) \, d\xi \, d\eta \quad (2.51)$$

For the line integration, the isoparametric mapping as shown in Figure 2.12 is used. In the physical element, the infinitesimal line segment $d\Gamma$ is mapped to $d\xi$ in the local coordinate as given by Equation (2.52). For

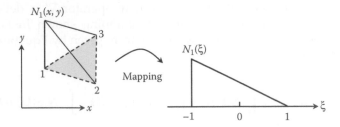

*Figure 2.12* Mapping for line integration in a 2D framework.

a linear element in $\xi$ coordinate, the shape functions are given by Equation (2.53) and the isoparametric mapping is given by Equation (2.54).

$$d\Gamma = \sqrt{dx^2 + dy^2} = \sqrt{\left(\frac{\partial x}{\partial \xi}\right)^2 + \left(\frac{\partial y}{\partial \xi}\right)^2} \, d\xi \qquad (2.52)$$

$$\left.\begin{aligned} N_1(\xi) &= 0.5(1 - \xi) \\ N_2(\xi) &= 0.5(1 + \xi) \end{aligned}\right\} \qquad (2.53)$$

$$\left.\begin{aligned} x &= N_1(\xi)x_1 + N_2(\xi)x_2 \\ y &= N_1(\xi)y_1 + N_2(\xi)y_2 \end{aligned}\right\} \qquad (2.54)$$

The infinitesimal line segment along line 1–2, $d\Gamma_{12}$, is given by Equation (2.55), where $L_{12}$ is the length of line 1–2. The line integral with numerical flux is evaluated as shown in Equation (2.56), where $\tilde{\mathbf{F}}_{12}$ is the numerical flux across line 1–2. For test functions $N_2$ and $N_3$ the numerical fluxes are given by Equation (2.57). In succinct notation, the line integral is given by Equation (2.58).

$$d\Gamma_{12} = \frac{L_{12}}{2} d\xi \qquad (2.55)$$

$$\int_{\Gamma_e} N_1(x,y)\tilde{\mathbf{F}} \, d\Gamma = \int_{-1}^{1} N_1(\xi)\tilde{\mathbf{F}}_{12} \frac{L_{12}}{2} d\xi + \int_{-1}^{1} N_1(\xi)\tilde{\mathbf{F}}_{31} \frac{L_{31}}{2} d\xi \qquad (2.56)$$

$$= \tilde{\mathbf{F}}_{12} \frac{L_{12}}{2} + \tilde{\mathbf{F}}_{31} \frac{L_{31}}{2}$$

$$\int_{\Gamma_e} N_2 \tilde{\mathbf{F}} \, d\Gamma = \tilde{\mathbf{F}}_{12} \frac{L_{12}}{2} + \tilde{\mathbf{F}}_{23} \frac{L_{23}}{2}, \quad \int_{\Gamma_e} N_3 \tilde{\mathbf{F}} \, d\Gamma = \tilde{\mathbf{F}}_{23} \frac{L_{23}}{2} + \tilde{\mathbf{F}}_{31} \frac{L_{31}}{2} \quad (2.57)$$

$$\int_{\Gamma_e} N_i \tilde{\mathbf{F}} \, d\Gamma = \sum_{j=1, j \neq i}^{3} \tilde{\mathbf{F}}_{ij} \frac{L_{ij}}{2} \quad (2.58)$$

## 2.4   Numerical integration

After transformation of the coordinates from global to local, the equations need to be integrated. For some simple cases, analytical integration is possible. However, for complex cases and more practically in computer implementations, the integrals are often evaluated numerically. Gaussian quadrature rules provide an easy way for numerical integration in the transformed or local coordinates system.

### 2.4.1   1D Gaussian quadrature

The 1D integral can be computed by the Gaussian quadrature formula in the interval [–1, 1] as shown in Equation (2.59), where $n_g$ is the number of quadrature points, $\xi_i$, $i = 1, 2, \ldots, n_g$ are the quadrature points, and $w_i$, $i = 1, 2, \ldots, n_g$ are the associated weights. For Gaussian quadrature points $n_g$, a polynomial function up to order $P = 2n_g - 1$ can be integrated exactly. The Gaussian quadrature rules, in the interval [–1, 1], for polynomial up to order 5 are listed in Table 2.1 (Reddy, 1993).

$$\int_{-1}^{1} f(\xi) \, d\xi \approx \sum_{i=1}^{n_g} w_i f(\xi_i) \quad (2.59)$$

**Table 2.1** Gaussian Quadrature for a Line Integral with the Interval [–1, 1]

| Number of Gaussian Points ($n_g$) | Gaussian Point Abscissae ($\xi$) | Weights ($w$) | Degree of Precision ($P$) |
|:---:|:---:|:---:|:---:|
| 1 | $\xi_1 = 0$ | $w_1 = 2$ | 1 |
| 2 | $\xi_1 = -1/\sqrt{3}$ $\xi_2 = 1/\sqrt{3}$ | $w_1 = w_2 = 1$ | 3 |
| 3 | $\xi_1 = -\sqrt{3}/\sqrt{5}$ $\xi_2 = 0$ $\xi_3 = \sqrt{3}/\sqrt{5}$ | $w_1 = 5/9$ $w_2 = 8/9$ $w_3 = 5/9$ | 5 |

**Table 2.2** Gaussian Quadrature and Weights for a Canonical Triangular Element

| Number of Gaussian Points ($n_g$) | Coordinates ($\xi, \eta$) | Weights ($w$) | Degree of Precision ($P$) |
|---|---|---|---|
| 1 | $(\xi_1, \eta_1) = \left(\dfrac{1}{3}, \dfrac{1}{3}\right)$ | $w_1 = 1$ | 1 |
| 3 | $(\xi_1, \eta_1) = (1/6, 1/6)$<br>$(\xi_2, \eta_2) = (4/6, 1/6)$<br>$(\xi_3, \eta_3) = (1/6, 4/6)$ | $w_1 = w_2 = w_3 = \dfrac{1}{3}$ | 2 |
| 7 | $a_1 = 0.1012865073235$<br>$a_2 = 0.7974269853531$<br>$a_3 = 0.4701420641051$<br>$a_4 = 0.0597158717898$<br>$(\xi_1, \eta_1) = (a_1, a_1)$<br>$(\xi_2, \eta_2) = (a_2, a_1)$<br>$(\xi_3, \eta_3) = (a_1, a_2)$<br>$(\xi_4, \eta_4) = (a_3, a_4)$<br>$(\xi_5, \eta_5) = (a_3, a_3)$<br>$(\xi_6, \eta_6) = (a_4, a_3)$<br>$(\xi_7, \eta_7) = (1/3, 1/3)$ | $w_1 = 0.1259391805$<br>$w_1 = w_2 = w_3$<br>$w_4 = 0.1323941527$<br>$w_4 = w_5 = w_6$<br>$w_7 = 0.225$ | 5 |

### 2.4.2   2D Gaussian quadrature for triangular elements

The 2D quadrature rule for the canonical triangular element is given by Equation (2.60) and the Gaussian quadrature points are listed in Table 2.2 (Reddy, 1993).

$$\int_0^1 \int_0^1 f(\xi, \eta) d\xi d\eta \approx \frac{1}{2} \sum_{i=1}^{n_g} w_i f(\xi_i, \eta_i) \qquad (2.60)$$

## 2.5   Approximate Riemann solvers

Since discontinuities are allowed across element boundaries, the solution for the numerical fluxes can be considered as a Riemann problem.

The Riemann problem for one-dimensional conservation laws is given by Equation (2.61), where the intercell flux $F(x = 0, t)$ needs to be solved.

$$
\left.
\begin{aligned}
&U_t + F(U)_x = 0 \\
&U(x,0) = U_L \quad \text{if} \quad x \leq 0 \\
&U(x,0) = U_R \quad \text{if} \quad x > 0
\end{aligned}
\right\} \tag{2.61}
$$

In this section, the HLL flux, HLLC flux, and Roe flux will be discussed in general form. Application of the Riemann solver to specific problems will be presented in the following chapters. More discussions and details about the Riemann problem can be found in Toro (2009).

### 2.5.1   HLL flux

Harten, Lax, and van Leer (1983) put forward the HLL Riemann solver with three constant states separated by two waves as shown in Figure 2.13. The slowest and fastest wave velocities $S_L$ and $S_R$ are assumed to be known. In the nontrivial case of $S_L < 0 < S_R$, the solution of numerical flux can be obtained through the integral relations for the control volume shown in Figure 2.14.

The conservation laws in the $x$–$t$ plane can be integrated as shown in Equation (2.62). After applying Green's theorem, the area integral can be reduced to a line integral as shown in Equation (2.63). Integrating Equation (2.63) over the left control volume $[TS_L, 0] \times [0, T]$ results in Equation (2.64). In the same way, integrating the conservation laws over the right control volume $[0, TS_R] \times [0, T]$ gives Equation (2.65). In these equations $F_L$, $F_R$, and $F_*$ are numerical fluxes across the element boundary.

$$
\iint (U_t + F_x)\,dx\,dt = 0 \tag{2.62}
$$

$$
\int_\Gamma U\,dx - F\,dt = 0 \tag{2.63}
$$

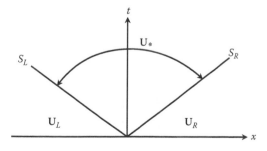

**Figure 2.13** Wave structure for the HLL Riemann solver.

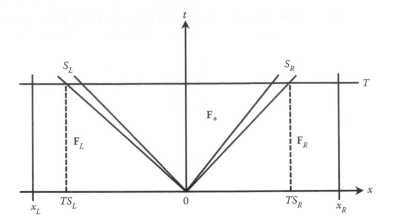

*Figure 2.14* Control volume for the HLL solver's wave structure.

$$(\mathbf{U}_L - \mathbf{U}_*)(0 - TS_L) - T(\mathbf{F}_* - \mathbf{F}_L) = 0 \tag{2.64}$$

$$(\mathbf{U}_R - \mathbf{U}_*)(TS_R - 0) - T(\mathbf{F}_R - \mathbf{F}_*) = 0 \tag{2.65}$$

Equations (2.64) and (2.65) can be solved to obtain the value of $\mathbf{F}_*$ as given by Equation (2.66). The trivial cases of $S_L \geq 0$ and $S_R \leq 0$ can be observed from the wave configuration. The HLL flux for the approximate Riemann problem is given by Equation (2.67). For the scalar case, where there is only one wave speed, $S = S_L = S_R$, the HLL flux will result in the upwind flux shown in Equation (2.68).

$$\mathbf{F}_* = \frac{S_R \mathbf{F}_L - S_L \mathbf{F}_R + S_L S_R (\mathbf{U}_R - \mathbf{U}_L)}{S_R - S_L} \tag{2.66}$$

$$\mathbf{F}^{HLL} = \begin{cases} \mathbf{F}_L & \text{if} & S_L \geq 0 \\ \dfrac{S_R \mathbf{F}_L - S_L \mathbf{F}_R + S_L S_R (\mathbf{U}_R - \mathbf{U}_L)}{S_R - S_L} & \text{if} & S_L < 0 < S_R \\ \mathbf{F}_R & \text{if} & S_R \leq 0 \end{cases} \tag{2.67}$$

$$F^{up} = \begin{cases} F_L & \text{if} & S \geq 0 \\ F_R & \text{if} & S < 0 \end{cases} \tag{2.68}$$

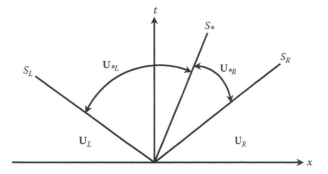

**Figure 2.15** HLLC Riemann solver's wave structure.

### 2.5.2 HLLC flux

The HLLC flux (Toro et al., 1989) is a modification of the HLL Riemann solver, where a three-wave structure is assumed as shown in Figure 2.15. Assuming the wave speeds $S_L$, $S_R$, and $S_*$ are known, the HLLC flux is given by Equation (2.69).

$$\mathbf{F}^{HLLC} = \begin{cases} \mathbf{F}(\mathbf{U}_L) & \text{if} & S_L \geq 0 \\ \mathbf{F}(\mathbf{U}_L) + S_L(\mathbf{U}_{*L} - \mathbf{U}_L) & \text{if} & S_L < 0 \leq S_* \\ \mathbf{F}(\mathbf{U}_R) + S_R(\mathbf{U}_{*R} - \mathbf{U}_R) & \text{if} & S_* < 0 < S_R \\ \mathbf{F}(\mathbf{U}_R) & \text{if} & S_R \leq 0 \end{cases} \tag{2.69}$$

### 2.5.3 Roe flux

The conservation laws in a general form can be written as shown in Equation (2.70), where the Jacobian matrix is given by Equation (2.71). In Roe's approach (Roe, 1981), the original Jacobian matrix $\mathbf{A}$ is replaced by a constant matrix $\tilde{\mathbf{A}}$ (which represents the local conditions) as shown in Equation (2.72). The original nonlinear system of conservation laws is converted into a linear system with constant coefficients as given by Equation (2.73).

$$\mathbf{U}_t + \mathbf{A}(\mathbf{U})\mathbf{U}_x = 0 \tag{2.70}$$

$$\mathbf{A}(\mathbf{U}) = \frac{\partial \mathbf{F}}{\partial \mathbf{U}} \tag{2.71}$$

$$\tilde{\mathbf{A}} = \mathbf{A}(\mathbf{U}_L, \mathbf{U}_R) \tag{2.72}$$

$$\mathbf{U}_t + \tilde{\mathbf{A}}\mathbf{U}_x = \mathbf{U}_t + \tilde{\mathbf{F}}_x = 0 \tag{2.73}$$

The solution to the linear system is given by Equation (2.74), where $m$ is the number of eigenvalues of the conservation law, $\tilde{\alpha}_i = \tilde{\alpha}_i(\mathbf{U}_L, \mathbf{U}_R)$ are the wave strengths, and $\tilde{\mathbf{K}}_i$ are the eigenvectors. Applying Green's theorem over a control volume (as in the case of HLL solver) results in the numerical flux for the original conservation law and is given by Equation (2.75). The numerical flux for the linearized system is given by Equation (2.76) based on the assumptions shown in Equation (2.77). The components in Equation (2.75) are then combined to determine the Roe numerical flux across the boundary as given by Equation (2.78).

$$\mathbf{U}_R - \mathbf{U}_L = \sum_{i=1}^{m} \tilde{\alpha}_i \tilde{\mathbf{K}}_i \tag{2.74}$$

$$\left. \begin{array}{l} \mathbf{F}_* = \mathbf{F}_L - S_L \mathbf{U}_L - \dfrac{1}{T} \displaystyle\int_{TS_L}^{0} \mathbf{U}\,dx \\[4mm] \mathbf{F}_* = \mathbf{F}_R - S_R \mathbf{U}_R + \dfrac{1}{T} \displaystyle\int_{0}^{TS_R} \mathbf{U}\,dx \end{array} \right\} \tag{2.75}$$

$$\left. \begin{array}{l} \tilde{\mathbf{F}}_* = \tilde{\mathbf{F}}_L - S_L \mathbf{U}_L - \dfrac{1}{T} \displaystyle\int_{TS_L}^{0} \tilde{\mathbf{U}}\,dx \\[4mm] \tilde{\mathbf{F}}_* = \tilde{\mathbf{F}}_R - S_R \mathbf{U}_R + \dfrac{1}{T} \displaystyle\int_{0}^{TS_R} \tilde{\mathbf{U}}\,dx \end{array} \right\} \tag{2.76}$$

$$\left. \begin{array}{c} \displaystyle\int_{TS_L}^{0} \mathbf{U}\,dx = \displaystyle\int_{TS_L}^{0} \tilde{\mathbf{U}}\,dx \\[4mm] \displaystyle\int_{0}^{TS_R} \mathbf{U}\,dx = \displaystyle\int_{0}^{TS_R} \tilde{\mathbf{U}}\,dx \\[4mm] \tilde{\mathbf{F}}(\mathbf{U}) = \tilde{\mathbf{A}}\mathbf{U} \end{array} \right\} \tag{2.77}$$

$$\mathbf{F}^{Roe} = \frac{1}{2}(\mathbf{F}_L + \mathbf{F}_R) - \frac{1}{2}\sum_{i=1}^{m} \tilde{\alpha}_i \,|\tilde{\lambda}_i|\, \tilde{\mathbf{K}}_i \tag{2.78}$$

Roe's Jacobian matrix is required to provide a linear mapping from the space $\mathbf{U}$ to $\mathbf{F}$. It must show consistency with exact Jacobian matrix as shown in Equation (2.79). In addition, it should preserve conservation across discontinuities by fulfilling the condition given by Equation (2.80). Last, the eigenvectors of $\tilde{\mathbf{A}}$ must be linearly independent.

$$\tilde{\mathbf{A}}(\mathbf{U}_L = \mathbf{U}_R = \mathbf{U}) = \mathbf{A}(\mathbf{U}) \tag{2.79}$$

$$\mathbf{F}(\mathbf{U}_R) - \mathbf{F}(\mathbf{U}_L) = \tilde{\mathbf{A}} \times (\mathbf{U}_R - \mathbf{U}_L) \tag{2.80}$$

## 2.6   Time integration

Since the discontinuous Galerkin method is a local formulation, the explicit time integration is preferred. The implicit time integration will result in a global matrix, which breaks the local formulation. For problems involving shock waves, the total variation diminishing (TVD) Runge–Kutta schemes are preferred. In this section, both the non-TVD and TVD Runge–Kutta time integration are described. In explicit schemes, the time step has to satisfy the CFL (Courant–Friedrichs–Lewy) condition for stability. More discussion will be given through examples in the following chapters. To illustrate the time integration procedure, Equation (2.81) will be used.

$$\frac{\partial \mathbf{U}}{\partial t} = \mathbf{M}^{-1}\mathbf{R} = \mathbf{L} \tag{2.81}$$

### 2.6.1   Non-TVD time integration

The time step is denoted by $\Delta t$, where $t^{n+1} = t^n + \Delta t$, $\mathbf{U}^n = \mathbf{U}(t = t^n)$, and $\mathbf{U}^{n+1} = \mathbf{U}(t = t^{n+1})$. The first-order Euler method is given by Equation (2.82). The non-TVD second-, third-, and fourth-order Runge–Kutta methods are given by Equations (2.83), (2.84), and (2.85), respectively (Li, 2006).

$$\mathbf{U}^{n+1} = \mathbf{U}^n + \Delta t \mathbf{L}(\mathbf{U}^n) \tag{2.82}$$

$$\left.\begin{aligned}
\mathbf{U}^{(1)} &= \mathbf{L}(\mathbf{U}^n) \\
\mathbf{U}^{(2)} &= \mathbf{L}(\mathbf{U}^n + \Delta t \mathbf{U}^{(1)}) \\
\mathbf{U}^{n+1} &= \mathbf{U}^n + \frac{\Delta t}{2}(\mathbf{U}^{(1)} + \mathbf{U}^{(2)})
\end{aligned}\right\} \tag{2.83}$$

$$U^{(1)} = L(U^n)$$

$$U^{(2)} = L\left(U^n + \frac{\Delta t}{2}U^{(1)}\right)$$

$$U^{(3)} = L(U^n - \Delta t U^{(1)} + 2\Delta t U^{(2)})$$

$$U^{n+1} = U^n + \frac{\Delta t}{6}(U^{(1)} + 4U^{(2)} + U^{(3)}) \tag{2.84}$$

$$U^{(1)} = L(U^n)$$

$$U^{(2)} = L\left(U^n + \frac{\Delta t}{2}U^{(1)}\right)$$

$$U^{(3)} = L\left(U^n + \frac{\Delta t}{2}U^{(2)}\right)$$

$$U^{(4)} = L(U^n + \Delta t U^{(3)})$$

$$U^{n+1} = U^n + \frac{\Delta t}{6}(U^{(1)} + 2U^{(2)} + 2U^{(3)} + U^{(4)}) \tag{2.85}$$

### 2.6.2  TVD time integration

The TVD Runge–Kutta time integration schemes are given by Gottlieb and Shu (1998). The second-, third-, and fourth-order TVD Runge–Kutta methods are given by Equations (2.86), (2.87), and (2.88), respectively.

$$U^{(1)} = U^n + \Delta t L(U^n)$$

$$U^{n+1} = \frac{1}{2}U^n + \frac{1}{2}U^{(1)} + \frac{\Delta t}{2}L(U^{(1)}) \tag{2.86}$$

$$U^{(1)} = U^n + \Delta t L(U^n)$$

$$U^{(2)} = \frac{3}{4}U^n + \frac{1}{4}U^{(1)} + \frac{\Delta t}{4}L(U^{(1)})$$

$$U^{n+1} = \frac{1}{3}U^n + \frac{2}{3}U^{(2)} + \frac{2}{3}\Delta t L(U^{(2)}) \tag{2.87}$$

$$\mathbf{U}^{(1)} = \mathbf{U}^n + \frac{\Delta t}{2}\mathbf{L}(\mathbf{U}^n)$$

$$\left.\begin{aligned}
\mathbf{U}^{(2)} &= \frac{649}{1600}\mathbf{U}^n - \frac{10890423}{25193600}\Delta t\mathbf{L}(\mathbf{U}^n) + \frac{951}{1600}\mathbf{U}^{(1)} + \frac{5000}{7873}\Delta t\mathbf{L}(\mathbf{U}^{(1)}) \\[2mm]
\mathbf{U}^{(3)} &= \frac{53989}{2500000}\mathbf{U}^n - \frac{102261}{5000000}\Delta t\mathbf{L}(\mathbf{U}^n) + \frac{4806213}{20000000}\mathbf{U}^{(1)} \\[2mm]
&\quad - \frac{5121}{20000}\Delta t\mathbf{L}(\mathbf{U}^{(1)}) + \frac{23619}{32000}\mathbf{U}^{(2)} + \frac{7873}{10000}\Delta t\mathbf{L}(\mathbf{U}^{(2)}) \\[2mm]
\mathbf{U}^{n+1} &= \frac{\mathbf{U}^n}{5} + \frac{\Delta t}{10}\mathbf{L}(\mathbf{U}^n) + \frac{6127\mathbf{U}^{(1)}}{30000} + \frac{\Delta t}{6}\mathbf{L}(\mathbf{U}^{(1)}) + \frac{7873\mathbf{U}^{(2)}}{30000} \\[2mm]
&\quad + \frac{\mathbf{U}^{(3)}}{3} + \frac{\Delta t}{6}\mathbf{L}(\mathbf{U}^{(3)})
\end{aligned}\right\} \qquad (2.88)$$

## References

Gottlieb, S., and Shu, C. W. (1998). Total variation diminishing Runge–Kutta schemes. *Mathematics of Computation*, 67(221), 73–85.

Harten, A., Lax, P. D., and van Leer, B. (1983). On upstream differencing and Godunov-type schemes for hyperbolic conservation laws. *SIAM Review*, 25(1), 35–61.

Laney, C. B. (1998). *Computational Gasdynamics*. Cambridge University Press, New York.

Lewis, R. W., Nithisrasu, P., and Seethearamu, K. N. (2004). *Fundamentals of the Finite Element Method for Heat and Fluid Flow*. Wiley, New York.

Li, B. Q. (2006). *Discontinuous Finite Elements in Fluid Dynamics and Heat Transfer*. Springer-Verlag, London.

Reddy, J. N. (1993). *An Introduction to the Finite Element Method*, 2nd ed. McGraw-Hill, New York.

Roe, P. (1981). Approximate Riemann solver, parameter vectors, and difference schemes. *Journal of Computational Physics*, 43(2), 357–372.

Toro, E. F. (2009). *Riemann Solvers and Numerical Methods for Fluid Dynamics*, 3rd ed. Springer-Verlag, Berlin, Heidelberg.

Toro, E., Spruce, M., and Speares, W. (1989). Restoration of the contact surface in the HLL-Riemann solver. *Shock Waves*, 4(1), 25–34.

Zienkiewicz, O. C., Taylor, R. L., and Zhu, J. Z. (2005). *The Finite Element Method: Its Basis and Fundamentals*. Elsevier, Spain.

# chapter three

# Discontinuous Galerkin method for one-dimensional nonconservative equations

In this chapter, the discontinuous Galerkin (DG) method is used to solve one-dimensional problems, which will give the readers a first glance at this method. Through examples, the steps needed for coordinate transformation, numerical integration, and applications of boundary and initial conditions are illustrated. The examples considered include the ordinary differential equation (ODE), pure convection, and pure diffusion in one dimension.

## 3.1 Discontinuous Galerkin method for ordinary differential equations

The discontinuous Galerkin method is applied first to find the numerical solution of an ordinary differential equation (ODE) given by Equation (3.1). The analytical solution of the equation is given by Equation (3.2). Linear and quadratic elements are used to discretize the domain. The continuous finite elements and the discontinuous finite elements configurations are shown in Figures 3.1 and 3.2, respectively. Within an element, the variable $u$ is approximated with Lagrangian polynomials (shape functions) as given by Equation (3.3).

$$\frac{du}{dx} = 1, \quad u(x = 1) = 1, \quad x \in [1,3] \tag{3.1}$$

$$u(x) = x, \quad x \in [1,3] \tag{3.2}$$

$$u \approx \hat{u} = \sum N_j(x) u_j \tag{3.3}$$

As shown in Figures 3.1 and 3.2, in the continuous finite element the approximate variable $\hat{u}$ is forced to be continuous across the element boundaries, so there will be only one value at a node. While in the discontinuous method, $\hat{u}$ is allowed to be discontinuous across the

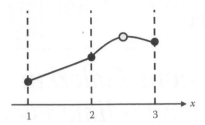

*Figure 3.1* Continuous elements.

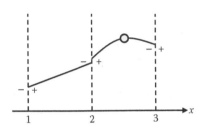

*Figure 3.2* Discontinuous elements.

boundaries, so $\hat{u}$ will have two different values defined on either sides of the boundary as shown in Equation (3.4). The approximation $\hat{u}$ is discontinuous only at the element boundaries, and the variation of $\hat{u}$ is smooth within an element. The ODE is multiplied by the test function $N_i(x)$ and the resulting equation is integrated over an element as shown in Equation (3.5), where $x_s^e$ and $x_e^e$ are the start and end coordinates of an element, respectively.

$$\left. \begin{aligned} u^- &= \lim_{x \uparrow x^-} \hat{u}(x) \\ u^+ &= \lim_{x \downarrow x^+} \hat{u}(x) \end{aligned} \right\} \tag{3.4}$$

$$\int_{x_s^e}^{x_e^e} N_i(x) \frac{dN_j(x)}{dx} u_j \, dx = \int_{x_s^e}^{x_e^e} N_i(x) dx \tag{3.5}$$

Integrating by parts, Equation (3.5) results in Equation (3.6), where $\tilde{u}$ is the numerical flux at element boundaries. Implementation of the isoparametric mapping results in Equation (3.7). For the linear element, $x \in [1, 2]$, the notations shown in Equation (3.8) are adopted, where the coordinates at the beginning and end of the linear element are given by Equation (3.9). The variations within the element are given by Equation (3.10), and the shape functions are given by Equation (3.11).

$$N_i(x)\tilde{u}\Big|_{x_s^e}^{x_e^e} - \left(\int_{x_s^e}^{x_e^e} \frac{dN_i(x)}{dx} N_j(x)dx\right) u_j = \int_{x_s^e}^{x_e^e} N_i(x)dx \qquad (3.6)$$

$$N_i(\xi)\tilde{u}\Big|_{-1}^{1} - \left(\int_{-1}^{1} \frac{dN_i(\xi)}{d\xi} N_j(\xi)d\xi\right) u_j = \int_{-1}^{1} N_i(\xi)\frac{\partial x}{\partial \xi} d\xi \qquad (3.7)$$

$$\left.\begin{aligned} u_1 &= u(x = 1^+) \\ u_2 &= u(x = 2^-) \end{aligned}\right\} \qquad (3.8)$$

$$\left.\begin{aligned} x_1 &= 1 \\ x_2 &= 2 \end{aligned}\right\} \qquad (3.9)$$

$$\left.\begin{aligned} \hat{u} &= N_1(\xi)u_1 + N_2(\xi)u_2 \\ x &= N_1(\xi)x_1 + N_2(\xi)x_2 \end{aligned}\right\} \qquad (3.10)$$

$$\left.\begin{aligned} N_1(\xi) &= 0.5(1-\xi) \\ N_2(\xi) &= 0.5(1+\xi) \end{aligned}\right\} \qquad (3.11)$$

The Jacobian is given by Equation (3.12). Substituting the value of the Jacobian in Equation (3.7) results in Equation (3.13). After substituting the test and basis functions into Equation (3.13), the resulting equation can be written as Equation (3.14). The numerical flux is approximated using an upwind scheme as given by Equation (3.15), where the boundary conditions are based on the values to the left of nodes.

$$J^e(\xi) = \frac{\partial x}{\partial \xi} = \frac{\partial\left(\sum_{j=1}^{2} N_j(\xi)x_j\right)}{\partial \xi} = \sum_{j=1}^{2} \frac{\partial N_j(\xi)}{\partial \xi} x_j = \frac{x_2 - x_1}{2} = \frac{\Delta x}{2} \qquad (3.12)$$

$$N_i(\xi)\tilde{u}\Big|_{-1}^{1} - \left(\int_{-1}^{1} \frac{dN_i(\xi)}{d\xi} N_j(\xi)d\xi\right) u_j = \int_{-1}^{1} N_i(\xi)\frac{\Delta x}{2} d\xi \qquad (3.13)$$

$$
\begin{bmatrix} -\tilde{u}_1 \\ \tilde{u}_2 \end{bmatrix} - \begin{bmatrix} \int_{-1}^{1} \frac{-1}{2}\frac{1}{2}(1-\xi)\,d\xi & \int_{-1}^{1} \frac{-1}{2}\frac{1}{2}(1+\xi)\,d\xi \\ \int_{-1}^{1} \frac{1}{2}\frac{1}{2}(1-\xi)\,d\xi & \int_{-1}^{1} \frac{1}{2}\frac{1}{2}(1+\xi)\,d\xi \end{bmatrix} \begin{bmatrix} u_1 \\ u_2 \end{bmatrix}
$$

$$
= \frac{\Delta x}{2} \begin{bmatrix} \int_{-1}^{1} \frac{1}{2}(1-\xi)\,d\xi \\ \int_{-1}^{1} \frac{1}{2}(1+\xi)\,d\xi \end{bmatrix}
\tag{3.14}
$$

$$
\left.\begin{aligned}
\tilde{u}_1 &= u(x = 1^-) = 1 \\
\tilde{u}_2 &= u(x = 2^-) = u_2
\end{aligned}\right\}
\tag{3.15}
$$

The integrals in Equation (3.14) can be evaluated analytically or numerically, and the resulting equation can be written as Equation (3.16). The equation can be simplified as shown in Equation (3.17). The solution of the equation is given by Equation (3.18), which is also the exact solution.

$$
\begin{bmatrix} -1 \\ u_2 \end{bmatrix} - \begin{bmatrix} -\dfrac{1}{2} & -\dfrac{1}{2} \\ \dfrac{1}{2} & \dfrac{1}{2} \end{bmatrix} \begin{bmatrix} u_1 \\ u_2 \end{bmatrix} = \frac{\Delta x}{2} \begin{bmatrix} 1 \\ 1 \end{bmatrix}
\tag{3.16}
$$

$$
\begin{bmatrix} -\dfrac{1}{2} & -\dfrac{1}{2} \\ \dfrac{1}{2} & -\dfrac{1}{2} \end{bmatrix} \begin{bmatrix} u_1 \\ u_2 \end{bmatrix} = -\begin{bmatrix} \dfrac{3}{2} \\ \dfrac{1}{2} \end{bmatrix}
\tag{3.17}
$$

$$
\begin{bmatrix} u_1 \\ u_2 \end{bmatrix} = \begin{bmatrix} 1 \\ 2 \end{bmatrix}
\tag{3.18}
$$

For the second element, $x \in [2, 3]$, a quadratic element, the notations given by Equation (3.19) are used, where the coordinates at the midpoint and ends of the element are given by Equation (3.20). For an isoparametric element, the variations within the element are given by Equation (3.21), where basis functions are given by Equation (3.22). The Jacobian for the quadratic element is given by Equation (3.23).

$$
\left.\begin{aligned}
u_1 &= u(x = 2^+) \\
u_2 &= u(x = 2.5) \\
u_3 &= u(x = 3^-)
\end{aligned}\right\} \tag{3.19}
$$

$$
\left.\begin{aligned}
x_1 &= 2 \\
x_2 &= 2.5 \\
x_3 &= 3
\end{aligned}\right\} \tag{3.20}
$$

$$
\left.\begin{aligned}
\hat{u} &= N_1(\xi)u_1 + N_2(\xi)u_2 + N_3(\xi)u_3 \\
x &= N_1(\xi)x_1 + N_2(\xi)x_2 + N_3(\xi)x_3
\end{aligned}\right\} \tag{3.21}
$$

$$
\left.\begin{aligned}
N_1(\xi) &= \frac{1}{2}\xi(\xi - 1) \\
N_2(\xi) &= (1 + \xi)(1 - \xi) \\
N_3(\xi) &= \frac{1}{2}\xi(1 + \xi)
\end{aligned}\right\} \tag{3.22}
$$

$$
J^e(\xi) = \frac{\partial x}{\partial \xi} = \frac{\partial\left(\sum_{j=1}^{3} N_j(\xi)x_j\right)}{\partial \xi} = \sum_{j=1}^{3} \frac{\partial N_j(\xi)}{\partial \xi} x_j \tag{3.23}
$$

Substituting the expressions for the basis functions and the Jacobian in Equation (3.7) and integrating the resulting equation gives Equation (3.24). Here again, the numerical flux is defined with an upwind scheme as given by Equation (3.25), where the boundary conditions are based on the values to the left of nodes. The final expression for the quadratic element is given by Equation (3.26), and the solution is given by Equation (3.27), which matches the exact solution.

$$
\begin{bmatrix} -\tilde{u}_1 \\ 0 \\ \tilde{u}_3 \end{bmatrix} - \begin{bmatrix} -1/2 & -2/3 & 1/6 \\ 2/3 & 0 & -2/3 \\ -1/6 & 2/3 & 1/2 \end{bmatrix}\begin{bmatrix} u_1 \\ u_2 \\ u_3 \end{bmatrix} = \begin{bmatrix} 1/6 \\ 2/3 \\ 1/6 \end{bmatrix} \tag{3.24}
$$

$$\left.\begin{aligned} \tilde{u}_1 &= u(x = 2^-) = 2 \\ \tilde{u}_3 &= u(x = 3^-) = u_3 \end{aligned}\right\} \tag{3.25}$$

$$\begin{bmatrix} -1/2 & -2/3 & 1/6 \\ 2/3 & 0 & -2/3 \\ -1/6 & 2/3 & -1/2 \end{bmatrix} \begin{bmatrix} u_1 \\ u_2 \\ u_3 \end{bmatrix} = - \begin{bmatrix} 13/6 \\ 2/3 \\ 1/6 \end{bmatrix} \tag{3.26}$$

$$\begin{bmatrix} u_1 \\ u_2 \\ u_3 \end{bmatrix} = \begin{bmatrix} 2 \\ 2.5 \\ 3 \end{bmatrix} \tag{3.27}$$

## 3.2   1D Linear convection

In this section, the discontinuous Galerkin method is applied to model the one-dimensional linear convection problem as given by Equation (3.28). The one-dimensional domain is $x \in [0, 1]$, and is divided into $Ne$ linear elements with $Ne + 1$ nodes $(0 = x_1 < x_2 \cdots < x_{Ne+1} = 1)$. For demonstration purpose, only linear and uniform size elements are used in this example (Figure 3.3). However, the element size can be nonuniform and the elements may be of different orders in practice.

$$\left.\begin{aligned} \frac{\partial C}{\partial t} + u\frac{\partial C}{\partial x} &= 0, \quad u = \text{const}, \quad x \in [0, 1] \\ C(x,0) &= \begin{cases} \sin(10\pi x) & \text{if} \quad x \in [0, 0.1] \\ 0 & \text{if} \quad x \in [0.1, 1] \end{cases} \end{aligned}\right\} \tag{3.28}$$

The variable $C$ is interpolated with the Lagrangian polynomial within an element as given by Equation (3.29). The equation is then multiplied by the test function, $N_i(x)$, and the resulting equation is integrated over an element, $x \in [x_s^e, x_e^e]$, as shown in Equation (3.30). By introducing the

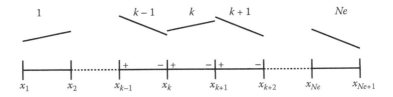

**Figure 3.3** One-dimensional domain with linear elements.

variation of a variable, Equation (3.31) is obtained. As $C_j$ is a function of time only, Equation (3.31) can be written as Equation (3.32). Using linear basis and test functions, and transforming the global coordinate to the local coordinate system results in Equation (3.33).

$$C \approx \hat{C} = \sum_{j=1}^{2} N_j C_j \tag{3.29}$$

$$\int_{x_s^e}^{x_e^e} N_i(x)\frac{\partial \hat{C}}{\partial t}dx + u\int_{x_s^e}^{x_e^e} N_i(x)\frac{\partial \hat{C}}{\partial x}dx = 0 \tag{3.30}$$

$$\int_{x_s^e}^{x_e^e} N_i(x)\frac{\partial N_j(x)C_j}{\partial t}dx +$$
$$u\left( N_i(x)\tilde{C}\Big|_{x_s^e}^{x_e^e} - \int_{x_s^e}^{x_e^e} \frac{\partial N_i(x)}{\partial x}N_j(x)C_j\,dx \right) = 0 \tag{3.31}$$

$$\int_{x_s^e}^{x_e^e} N_i(x)N_j(x)dx\frac{\partial C_j}{\partial t} + u N_i(x)\tilde{C}\Big|_{x_s^e}^{x_e^e} -$$
$$u\left( \int_{x_s^e}^{x_e^e} \frac{\partial N_i(x)}{\partial x}N_j(x)dx \right)C_j = 0 \tag{3.32}$$

$$\frac{\Delta x}{2}\int_{-1}^{1} N_i(\xi)N_j(\xi)d\xi\frac{\partial C_j}{\partial t} + u N_i(\xi)\tilde{C}\Big|_{-1}^{1} -$$
$$u\left( \int_{-1}^{1} \frac{\partial N_i(\xi)}{\partial \xi}N_j(\xi)d\xi \right)C_j = 0 \tag{3.33}$$

Integrating Equation (3.33) results in Equation (3.34), which can be simplified to Equation (3.35). The numerical flux at the boundary is approximated using an upwind scheme given by Equation (3.36). Finally, Equation (3.35) can be written in a simplified form as Equation (3.37).

$$\Delta x \begin{bmatrix} 2/6 & 1/6 \\ 1/6 & 2/6 \end{bmatrix}\frac{\partial}{\partial t}\begin{bmatrix} C_1 \\ C_2 \end{bmatrix} + u\begin{bmatrix} -\tilde{C}_1 \\ \tilde{C}_2 \end{bmatrix} - u\begin{bmatrix} -1/2 & -1/2 \\ 1/2 & 1/2 \end{bmatrix}\begin{bmatrix} C_1 \\ C_2 \end{bmatrix} = 0 \tag{3.34}$$

$$\frac{\partial}{\partial t}\begin{bmatrix} C_1 \\ C_2 \end{bmatrix} = \frac{u}{\Delta x}\begin{bmatrix} -3 & -3 \\ 3 & 3 \end{bmatrix}\begin{bmatrix} C_1 \\ C_2 \end{bmatrix} - \frac{u}{\Delta x}\begin{bmatrix} -4 & -2 \\ 2 & 4 \end{bmatrix}\begin{bmatrix} \tilde{C}_1 \\ \tilde{C}_2 \end{bmatrix} \tag{3.35}$$

$$\tilde{C}_j = \begin{cases} C_j^- & \text{if} \quad u > 0 \\ C_j^+ & \text{if} \quad u < 0 \end{cases} \tag{3.36}$$

$$\frac{\partial C}{\partial t} = L = \frac{u}{\Delta x} \begin{bmatrix} -3 & -3 \\ 3 & 3 \end{bmatrix} \begin{bmatrix} C_1 \\ C_2 \end{bmatrix} - \frac{u}{\Delta x} \begin{bmatrix} -4 & -2 \\ 2 & 4 \end{bmatrix} \begin{bmatrix} \tilde{C}_1 \\ \tilde{C}_2 \end{bmatrix} \tag{3.37}$$

Time integration has to be used to find the new values at the next time step. Former studies have shown that the total variation diminishing (TVD) Runge–Kutta time integration scheme should be one order higher than the polynomial used for space discretization (Cockburn and Lin, 1989; Cockburn and Shu, 1989; Cockburn et al., 1990). The second-order TVD Runge–Kutta scheme is used here for linear elements ($p = 1$) used in this example. For $u = 1$, the numerical results at $t = 0.6$ are sought. The effects of time step size and element size on numerical accuracy will be explored. Numerical results with 1000 elements and different time steps are shown in Figure 3.4. The Courant number is defined as $Cr = |u| \Delta t / \Delta x$. For the 1D convection problem with constant velocity $u$, the CFL (Courant–Friedrichs–Lewy) condition given by $Cr = (2p + 1)^{-1}$

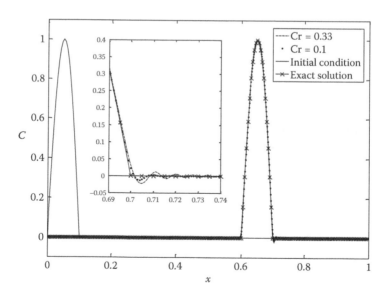

*Figure 3.4* Comparison of numerical (with different time steps) and analytical solutions for a 1D linear convection.

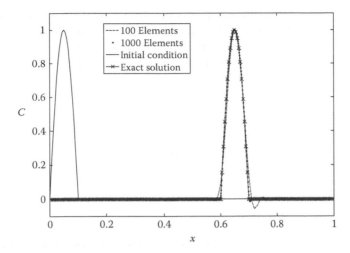

***Figure 3.5*** Comparison of numerical (with different spatial discretization) and analytical solutions for a 1D linear convection.

needs to be satisfied to ensure the stability (Cockburn, 1999), where $p$ is the order of the polynomial used for space discretization.

$$\left.\begin{array}{l} \mathbf{C}^{(1)} = \mathbf{C}^n + \Delta t \mathbf{L}(\mathbf{C}^n) \\[2mm] \mathbf{C}^{n+1} = \dfrac{1}{2}[\mathbf{C}^n + \mathbf{C}^{(1)} + \Delta t \mathbf{L}(\mathbf{C}^{(1)})] \end{array}\right\} \tag{3.38}$$

In Figure 3.4, numerical results for Courant numbers of 0.1 and 0.33 are shown. As the time step size is increased, oscillations are observed at the leading end of the wave. The numerical results with the Courant number of 0.1 are in good agreement with the exact solution. In Figure 3.5, the numerical results with 100 and 1000 elements for the Courant number of 0.1 are shown. It is obvious that with the refinement of mesh size, more accurate numerical results are obtained. The issues about numerical stability, convergence, consistency, and numerical errors are not discussed in this book. Interested readers can find more details about these topics in the books dealing with numerical methods or computational fluid dynamics (Hirsch, 1988, 1990; Chung, 2002; Li, 2006).

## 3.3   1D Transient diffusion

The one-dimensional transient linear diffusion problem with constant source term ($Q$) is solved using the discontinuous Galerkin method. The problem along with the boundary and initial conditions are given by

Equation (3.39), where $D$ is the diffusion coefficient. In the discontinuous Galerkin formulation, since both the shape functions and test functions are discontinuous across the element boundaries, the second-order spatial term needs to be treated in a mixed form with the first-order derivative (Li, 2006).

$$
\left.\begin{aligned}
\frac{\partial C}{\partial t} &= D\frac{\partial^2 C}{\partial x^2} + Q, \quad x \in [0,1] \\
D &= 1 \\
C(x=0) &= C(x=1) = 0 \\
C(x,t=0) &= 0
\end{aligned}\right\}
\tag{3.39}
$$

Integrating the second-order derivative term by parts, after multiplying by the test function, results in Equation (3.40). Difficulties arise when defining the flux of $\partial C/\partial x$ across the element boundaries. An intermediate variable associated with the first-order derivative is used to circumvent the problem. By defining an intermediate variable, as given by Equation (3.41), the diffusion problem can be written as Equation (3.42).

$$
\int_{x_s^e}^{x_e^e} N_i \frac{\partial^2 C}{\partial x^2} dx = N_i \frac{\partial C}{\partial x}\Big|_{x_s^e}^{x_e^e} - \int_{x_s^e}^{x_e^e} \frac{\partial N_i}{\partial x}\frac{\partial C}{\partial x} dx
\tag{3.40}
$$

$$
q = -\frac{\partial C}{\partial x} \quad \text{or} \quad q + \frac{\partial C}{\partial x} = 0
\tag{3.41}
$$

$$
\frac{\partial C}{\partial t} + \frac{\partial q}{\partial x} = Q
\tag{3.42}
$$

The domain $x \in [0, 1]$ is divided into $Ne$ linear elements with $Ne + 1$ nodes $(0 = x_1 < x_2 \cdots < x_{Ne+1} = 1)$. The variation of variables within an element is approximated as shown in Equation (3.43). Equation (3.41) is multiplied by test function $N_i(x)$ and then integrated over an element. The concentration gradient term is integrated by parts, the integral is transformed into a local coordinate system using isoparametric mapping, and finally the integration is performed. These steps are given by Equations (3.44) to (3.48).

$$
\left.\begin{aligned}
q \approx \hat{q} &= \sum N_j(x)q_j \\
C \approx \hat{C} &= \sum N_j(x)C_j
\end{aligned}\right\}
\tag{3.43}
$$

$$\int_{x_s^e}^{x_e^e} N_i(x)\hat{q}\,dx + \int_{x_s^e}^{x_e^e} N_i(x)\frac{\partial \hat{C}}{\partial x}\,dx = 0 \tag{3.44}$$

$$\left(\int_{x_s^e}^{x_e^e} N_i(x)N_j(x)dx\right)q_j + N_i(x)\tilde{C}\Big|_{x_s^e}^{x_e^e} - \left(\int_{x_s^e}^{x_e^e}\frac{\partial N_i(x)}{\partial x}N_j(x)dx\right)C_j = 0 \tag{3.45}$$

$$\frac{\Delta x}{2}\left(\int_{-1}^{1} N_i(\xi)N_j(\xi)d\xi\right)q_j + N_i(\xi)\tilde{C}\Big|_{-1}^{1} -$$
$$\left(\int_{-1}^{1}\frac{\partial N_i(\xi)}{\partial \xi}N_j(\xi)d\xi\right)C_j = 0 \tag{3.46}$$

$$\Delta x\begin{bmatrix} 2/6 & 1/6 \\ 1/6 & 2/6 \end{bmatrix}\begin{bmatrix} q_1 \\ q_2 \end{bmatrix} + \begin{bmatrix} -\tilde{C}_1 \\ \tilde{C}_2 \end{bmatrix} - \begin{bmatrix} -1/2 & -1/2 \\ 1/2 & 1/2 \end{bmatrix}\begin{bmatrix} C_1 \\ C_2 \end{bmatrix} = \begin{bmatrix} 0 \\ 0 \end{bmatrix} \tag{3.47}$$

$$\begin{bmatrix} q_1 \\ q_2 \end{bmatrix} = \frac{1}{\Delta x}\begin{bmatrix} -3 & -3 \\ 3 & 3 \end{bmatrix}\begin{bmatrix} C_1 \\ C_2 \end{bmatrix} - \frac{1}{\Delta x}\begin{bmatrix} -4 & 2 \\ -2 & 4 \end{bmatrix}\begin{bmatrix} \tilde{C}_1 \\ \tilde{C}_2 \end{bmatrix} \tag{3.48}$$

A similar procedure as described above for Equation (3.41) is adopted for Equation (3.42), and the results are given by Equations (3.49) to (3.53). In Equation (3.54), the numerical fluxes are calculated using the center flux for this diffusion problem. In this example, the source term ($Q$) is taken as 1.0.

$$\int_{x_s^e}^{x_e^e} N_i(x)\frac{\partial \hat{C}}{\partial t}\,dx + \int_{x_s^e}^{x_e^e} N_i(x)\frac{\partial \hat{q}}{\partial x}\,dx = Q\int_{x_s^e}^{x_e^e} N_i(x)dx \tag{3.49}$$

$$\left(\int_{x_s^e}^{x_e^e} N_i(x)N_j(x)dx\right)\frac{\partial C_j}{\partial t} + N_i(x)\tilde{q}\Big|_{x_s^e}^{x_e^e} -$$
$$\left(\int_{x_s^e}^{x_e^e}\frac{\partial N_i(x)}{\partial x}N_j(x)dx\right)q_j = Q\int_{x_s^e}^{x_e^e} N_i(x)dx \tag{3.50}$$

$$\frac{\Delta x}{2}\left(\int_{-1}^{1} N_i(\xi)N_j(\xi)d\xi\right)\frac{\partial C_j}{\partial t} + N_i(\xi)\tilde{q}\Big|_{-1}^{1} -$$
$$\left(\int_{-1}^{1}\frac{\partial N_i(\xi)}{\partial \xi}N_j(\xi)d\xi\right)q_j = Q\frac{\Delta x}{2}\int_{-1}^{1} N_i(\xi)d\xi \tag{3.51}$$

$$\Delta x \begin{bmatrix} 2/6 & 1/6 \\ 1/6 & 2/6 \end{bmatrix} \frac{\partial}{\partial t} \begin{bmatrix} C_1 \\ C_2 \end{bmatrix} + \begin{bmatrix} -\tilde{q}_1 \\ \tilde{q}_2 \end{bmatrix} - \begin{bmatrix} -1/2 & -1/2 \\ 1/2 & 1/2 \end{bmatrix} \begin{bmatrix} q_1 \\ q_2 \end{bmatrix} = \frac{Q\Delta x}{2} \begin{bmatrix} 1 \\ 1 \end{bmatrix} \quad (3.52)$$

$$\frac{\partial}{\partial t} \begin{bmatrix} C_1 \\ C_2 \end{bmatrix} = \frac{1}{\Delta x} \begin{bmatrix} -3 & -3 \\ 3 & 3 \end{bmatrix} \begin{bmatrix} q_1 \\ q_2 \end{bmatrix} + \frac{1}{\Delta x} \begin{bmatrix} 4 & -2 \\ -2 & 4 \end{bmatrix} \begin{bmatrix} Q\Delta x/2 + \tilde{q}_1 \\ Q\Delta x/2 - \tilde{q}_2 \end{bmatrix} \quad (3.53)$$

$$\left. \begin{aligned} \tilde{C}_k &= \frac{C_k^- + C_k^+}{2} \\ \tilde{q}_k &= \frac{q_k^- + q_k^+}{2} \end{aligned} \right\} \quad (3.54)$$

The computational procedure consists of first enforcing the initial condition $C(x, t = 0)$. To proceed the computation from a time step $n$ to a time step $n + 1$, the unknowns $q^n$ are first solved using Equation (3.48), then $C^{n+1}$ is calculated using Equation (3.53) with proper time integration, and finally $C^{n+1}$ is substituted back into Equation (3.48) to calculate $q^{n+1}$. This procedure continues until the time reaches the specified time or a steady state is reached. The first-order Euler forward scheme is adopted for time integration. Defining the diffusion number as $d = D\Delta t/\Delta x^2$, numerical results show that for linear elements, as used in this example, the diffusion number should be less than or equal to 0.125 in order to achieve stable solutions. The numerical results with 200 elements, that is, $\Delta x = 0.005$, are shown in Figure 3.6.

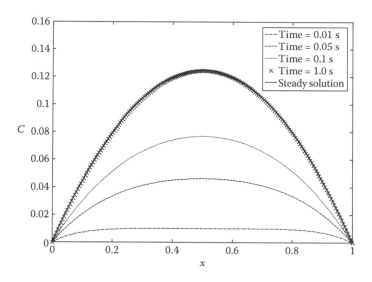

*Figure 3.6* Numerical solutions at different times for a 1D unsteady diffusion.

## 3.4   1D Steady diffusion

The governing equation of 1D steady diffusion, Equation (3.55), is solved using the discontinuous Galerkin method. To apply the discontinuous Galerkin method, the governing equation can be written in the mixed form as given by Equation (3.56) or Equation (3.57). The later form is widely used by researchers (Castillo et al., 2001; Li, 2006). Numerical results show that the choice of the mixed form is important. Here, Equation (3.57) is adopted. The original diffusion equation can be written as Equations (3.58) and (3.59).

$$\frac{d^2C}{dx^2} + Q = 0 \tag{3.55}$$

$$\left.\begin{aligned} q &= -\frac{dC}{dx} \\[2mm] -\frac{dq}{dx} + Q &= 0 \end{aligned}\right\} \tag{3.56}$$

$$\left.\begin{aligned} q &= \frac{dC}{dx} \\[2mm] \frac{dq}{dx} + Q &= 0 \end{aligned}\right\} \tag{3.57}$$

$$\frac{dC}{dx} - q = 0, \quad x \in [0,1] \tag{3.58}$$

$$\frac{dq}{dx} + Q = 0, \quad x \in [0,1] \tag{3.59}$$

The boundary conditions of $C(x = 0) = C(x = 1) = 0$ and $Q = 1.0$ are used in the simulation. The discontinuous Galerkin procedure is similar to that used before. The domain $x \in [0, 1]$ is divided into $Ne$ linear elements with $Ne + 1$ nodes $(0 = x_1 < x_2 \cdots < x_{Ne+1} = 1)$. Applying the procedure discussed in the previous section, Equation (3.58) results in Equations (3.60) and (3.61), while Equation (3.59) is transformed into Equations (3.62) and (3.63). Equations (3.61) and (3.63) can be combined to obtain Equation (3.64).

$$N_i(\xi)\tilde{C}\Big|_{-1}^{1} - \left(\int_{-1}^{1} \frac{\partial N_i(\xi)}{\partial \xi} N_j(\xi)d\xi\right)C_j - \\ \frac{\Delta x}{2}\left(\int_{-1}^{1} N_i(\xi)N_j(\xi)d\xi\right)q_j = 0 \tag{3.60}$$

$$\begin{bmatrix} -\tilde{C}_1 \\ \tilde{C}_2 \end{bmatrix} - \begin{bmatrix} -1/2 & -1/2 \\ 1/2 & 1/2 \end{bmatrix} \begin{bmatrix} C_1 \\ C_2 \end{bmatrix} - \Delta x \begin{bmatrix} 2/6 & 1/6 \\ 1/6 & 2/6 \end{bmatrix} \begin{bmatrix} q_1 \\ q_2 \end{bmatrix} = 0 \quad (3.61)$$

$$N_i(\xi)\tilde{q}\Big|_{-1}^{1} - \left( \int_{-1}^{1} \frac{\partial N_i(\xi)}{\partial \xi} N_j(\xi) d\xi \right) q_j + Q\frac{\Delta x}{2} \int_{-1}^{1} N_i(\xi) d\xi = 0 \quad (3.62)$$

$$\begin{bmatrix} -\tilde{q}_1 \\ \tilde{q}_2 \end{bmatrix} - \begin{bmatrix} -1/2 & -1/2 \\ 1/2 & 1/2 \end{bmatrix} \begin{bmatrix} q_1 \\ q_2 \end{bmatrix} + \frac{Q\Delta x}{2} \begin{bmatrix} 1 \\ 1 \end{bmatrix} = 0 \quad (3.63)$$

$$\begin{bmatrix} -1/2 & -1/2 & \Delta x/3 & \Delta x/6 \\ 1/2 & 1/2 & \Delta x/6 & \Delta x/3 \\ 0 & 0 & -1/2 & -1/2 \\ 0 & 0 & 1/2 & 1/2 \end{bmatrix} \begin{bmatrix} C_1 \\ C_2 \\ q_1 \\ q_2 \end{bmatrix} = \begin{bmatrix} -\tilde{C}_1 \\ \tilde{C}_2 \\ -\tilde{q}_1 + Q\Delta x/2 \\ \tilde{q}_2 + Q\Delta x/2 \end{bmatrix} \quad (3.64)$$

The 4 × 4 matrix, given by Equation (3.64), is singular. In order to solve this system, a proper choice of numerical flux is required. In addition, the numerical flux should meet the stability requirement, and satisfy the existence and uniqueness conditions. Discussion about the numerical flux for steady-state heat conduction problems can be found in literature (Castillo et al., 2001; Arnold et al., 2002; Li, 2006). In this example, the numerical fluxes are approximated as given by Equation (3.65). Substituting the numerical flux functions into Equation (3.64) results in Equation (3.66), where $b_1 = a_{12} + 0.5$ and $b_2 = a_{12} - 0.5$.

$$\left. \begin{aligned} \tilde{C}_k &= \frac{1}{2}(C_k^- + C_k^+) + a_{12}(C_k^- - C_k^+) \\ \tilde{q}_k &= \frac{1}{2}(q_k^- + q_k^+) - a_{11}(C_k^- - C_k^+) - a_{12}(q_k^- - q_k^+) \end{aligned} \right\} \quad (3.65)$$

$$\begin{bmatrix} -a_{12} & -0.5 & \Delta x/3 & \Delta x/6 \\ 0.5 & -a_{12} & \Delta x/6 & \Delta x/3 \\ a_{11} & 0 & a_{12} & -0.5 \\ 0 & a_{11} & 0.5 & a_{12} \end{bmatrix} \begin{bmatrix} C_1 \\ C_2 \\ q_1 \\ q_2 \end{bmatrix} = \begin{bmatrix} -b_1 C_1^- \\ -b_2 C_2^+ \\ b_2 q_1^- + a_{11} C_1^- + Q\Delta x/2 \\ b_1 q_2^+ + a_{11} C_2^+ + Q\Delta x/2 \end{bmatrix} \quad (3.66)$$

Equation (3.66) can be solved using an iterative method. The unknowns $C$ and $q$ are initially set to zero. Boundary conditions are enforced before an iteration $n$ ($n = 1, 2, ...$). During the iteration $n$, calculations are performed in order, from the first element to the last element. In an element $e$, $C_1^-$ and $q_1^-$ are calculated from the preceding element or initial guess and $C_2^+$ and $q_2^+$ are calculated from the following element or initial guess. These values are substituted into the right-hand side of Equation (3.66), and new values at the iteration $n$ are obtained. The calculations are carried on with the updated data until the convergence criterion ($\varepsilon$) is met or stopped at a specific iteration number. The $L^2$-norm based on the values at iteration levels $n + 1$ and $n$, as given by Equation (3.67), can be used to determine the convergence of the solution.

$$
\left[ \frac{\displaystyle\sum_{j=1}^{Ne+1}\left(\left(\Delta C_j^-\right)^2 + \left(\Delta C_j^+\right)^2 + \left(\Delta q_j^-\right)^2 + \left(\Delta q_j^+\right)^2\right)}{\displaystyle\sum_{j=1}^{Ne+1}\left(C_{j,n+1}^{-}{}^2 + C_{j,n+1}^{+}{}^2 + q_{j,n+1}^{-}{}^2 + q_{j,n+1}^{+}{}^2\right)} \right] < \varepsilon \qquad (3.67)
$$

Castillo et al. (2001) show that $a_{11} > 0$ is required to provide a stable, convergent, and unique solution. In this case, $a_{11} = 1$, and $a_{12} = 0.5, -0.5$, and 0 for upwind, downwind, and central flux for $C$, respectively, are used. Numerical results with 10 elements are shown in Figure 3.7. Results show

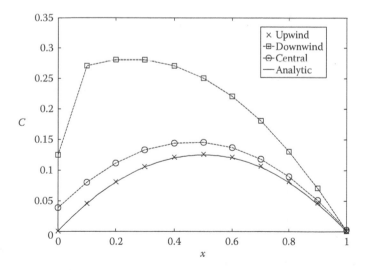

*Figure 3.7* Comparison of numerical (with different flux parameters) and analytical solutions for steady diffusion.

that the choice of parameters has a great influence on the performance of the numerical scheme.

As opposed to Equation (3.57), a mixed form given by Equation (3.56) can be used. The boundary flux functions as given by Equation (3.65) are used. Numerical results show that $a_{11} < 0$ is required to achieve a stable, convergent, and unique solution in this case. The resulting integrated equation is given by Equation (3.68).

$$
\begin{bmatrix}
-a_{12} & -0.5 & \Delta x/3 & \Delta x/6 \\
0.5 & -a_{12} & \Delta x/6 & \Delta x/3 \\
a_{11} & 0 & a_{12} & -0.5 \\
0 & a_{11} & 0.5 & a_{12}
\end{bmatrix}
\begin{bmatrix}
C_1 \\
C_2 \\
q_1 \\
q_2
\end{bmatrix}
=
\begin{bmatrix}
-b_1 C_1^- \\
-b_2 C_2^+ \\
b_2 q_1^- + a_{11} C_1^- + Q\Delta x/2 \\
b_1 q_2^+ + a_{11} C_2^+ + Q\Delta x/2
\end{bmatrix}
\tag{3.68}
$$

In the application of the discontinuous Galerkin method for elliptic problems, the choices of the mixed form and the boundary flux function parameters determine the numerical performance. However, with the continuous Galerkin finite element method, the elliptic problem with second-order spatial derivatives can be easily handled. So before the application of the discontinuous Galerkin method, the reader should have a clear understanding of the equations type. As the discontinuous Galerkin method is efficient and suitable to model hyperbolic equations in conservation forms, attention will be focused on these types of problems. Further details regarding the application of the discontinuous Galerkin method to elliptic and parabolic equations can be found in Rivière (2008).

## *References*

Arnold, D. N., Brezzi, F., Cockburn, B., and Marini, L. D. (2002). *Unified Analysis of Discontinuous Implementation.* SIAM, Philadelphia.

Castillo, P., Cockburn, B., Perugia, I., and Schötzau, D. (2001). An a priori error analysis of the local discontinuous Galerkin method for elliptic problems. *SIAM Journal on Numerical Analysis*, 38(5), 1676–1706.

Chung, T. J. (2002). *Computational Fluid Dynamics.* Cambridge University Press, Cambridge.

Cockburn, B. (1999). Discontinuous Galerkin methods for convection dominated problems. In *High-Order Methods for Computational Physics (Lecture Notes in Computational Science and Engineering*, vol. 9), edited by T. Barth and H. Deconinck, Springer-Verlag, New York, 69–224.

Cockburn, B., Hou, S., and Shu, C. W. (1990). The Runge–Kutta local projection discontinuous Galerkin finite element method for conservation laws IV: The multidimensional case. *Mathematics of Computation*, 54(190), 545–581.

Cockburn, B., and Lin, S. Y. (1989). TVB Runge–Kutta local projection discontinuous Galerkin finite element method for conservation laws III: One-dimensional systems. *Journal of Computational Physics*, 84(1), 90–113.

Cockburn, B., and Shu, C. W. (1989). TVB Runge–Kutta local projection discontinuous Galerkin finite element method for conservation laws II: General framework. *Mathematics of Computation*, 52(186), 411–435.

Hirsch, C. (1988). *Numerical Computation of Internal and External Flows, Volume 1: Fundamentals of Numerical Discretization.* Wiley, New York.

Hirsch, C. (1990). *Numerical Computation of Internal and External Flows, Volume 2: Computational Methods for Inviscid and Viscous Flows.* Wiley, New York.

Li, B. Q. (2006). *Discontinuous Finite Elements in Fluid Dynamics and Heat Transfer.* Springer-Verlag, London.

Rivière, B. (2008). *Discontinuous Galerkin Methods for Solving Elliptic and Parabolic Equations: Theory and Implementation.* SIAM, Philadelphia.

*chapter four*

# One-dimensional conservation laws

In this and the following chapters, the discontinuous Galerkin (DG) method will be applied to numerically solve the shallow water flow problems, which are governed by hyperbolic conservation laws. The scalar and system of conservation laws in one dimension are discussed in this chapter. The method is first applied to the benchmark Burgers' equation, followed by the application of the DG method to solve the shallow water flow problems in rectangular channels. The total variation diminishing (TVD) slope limiter is presented for the nonlinear conservation laws.

## 4.1  Burgers' equation

Burgers' equation, which represents a nonlinear convection problem, has been used as a benchmark test for evaluating numerical schemes. Burgers' equation is an idealized case for the shock wave and rarefaction wave phenomena in fluid flow. In this section, the discontinuous Galerkin method is applied to solve Burgers' equation.

### 4.1.1  Properties of Burgers' equation

Burgers' equation in conservation form and its characteristic wave speed (eigenvalue) are given by Equations (4.1) and (4.2), respectively. In Burgers' equation, the shock wave and rarefaction wave are generated depending on the initial conditions. The initial condition with discontinuity is given by Equation (4.3). The solution, $u(x, t)$, for $u_L > u_R$ and $u_L < u_R$, are given, respectively, by Equations (4.4) and (4.5). These solutions represent the shock wave and rarefaction wave, respectively.

$$\frac{\partial u}{\partial t} + \frac{\partial f(u)}{\partial x} = \frac{\partial u}{\partial t} + \frac{\partial 0.5u^2}{\partial x} = 0 \tag{4.1}$$

$$\lambda = \frac{\partial f(u)}{\partial u} = \frac{\partial 0.5u^2}{\partial u} = u \tag{4.2}$$

$$u(x,0) = \begin{cases} u_L & \text{if} \quad x < 0 \\ \dfrac{u_L + u_R}{2} & \text{if} \quad x = 0 \\ u_R & \text{if} \quad x > 0 \end{cases} \qquad (4.3)$$

$$u(x,t) = \begin{cases} u_L & \text{if} \quad \dfrac{x}{t} < \dfrac{(u_L + u_R)}{2} \\ \dfrac{u_L + u_R}{2} & \text{if} \quad \dfrac{x}{t} = \dfrac{(u_L + u_R)}{2} \\ u_R & \text{if} \quad \dfrac{x}{t} > \dfrac{(u_L + u_R)}{2} \end{cases} \qquad (4.4)$$

$$u(x,t) = \begin{cases} u_L & \text{if} \quad \dfrac{x}{t} \le u_L \\ \dfrac{x}{t} & \text{if} \quad u_L < \dfrac{x}{t} < u_R \\ u_R & \text{if} \quad \dfrac{x}{t} \ge u_R \end{cases} \qquad (4.5)$$

## 4.1.2   Discontinuous Galerkin formulation for Burgers' equation

Burgers' equation and the initial condition used in this example are given by Equation (4.6). The one-dimensional domain is divided into $Ne$ linear elements with $Ne + 1$ nodes $(-1 = x_1 < x_2 \cdots < x_{Ne+1} = 1)$. The variation of variables within an element are approximated as given by Equation (4.7).

$$\left. \begin{array}{c} \dfrac{\partial u}{\partial t} + \dfrac{\partial f(u)}{\partial x} = \dfrac{\partial u}{\partial t} + \dfrac{\partial 0.5u^2}{\partial x} = 0, \quad x \in [-1,1] \\[2mm] u(x,0) = \begin{cases} u_L, & x \in [-1,0) \\ u_R, & x \in (0,1] \end{cases} \end{array} \right\} \qquad (4.6)$$

$$\left. \begin{array}{c} u \approx \hat{u} = \sum N_j(x)u_j \\[2mm] f \approx \hat{f} = f(\hat{u}) = 0.5\hat{u}^2 \end{array} \right\} \qquad (4.7)$$

The governing equation is multiplied by the test function, $N_i(x)$, and integrated over an element $[x_s^e, x_e^e]$. The flux term is integrated by parts and the numerical flux at the boundaries of an element is denoted by $\tilde{f}$. The unknowns in the element are substituted with approximation given by Equation (4.7). Finally, the physical domain is transformed into a local coordinate system. These steps are given, respectively, by Equations (4.8) to (4.11).

$$\int_{x_s^e}^{x_e^e} N_i(x) \frac{\partial \hat{u}}{\partial t} dx + \int_{x_s^e}^{x_e^e} N_i(x) \frac{\partial \hat{f}}{\partial x} dx = 0 \tag{4.8}$$

$$\int_{x_s^e}^{x_e^e} N_i(x) \frac{\partial \hat{u}}{\partial t} dx + N_i(x)\tilde{f} \Big|_{x_s^e}^{x_e^e} - \int_{x_s^e}^{x_e^e} \frac{\partial N_i(x)}{\partial x} \hat{f} dx = 0 \tag{4.9}$$

$$\left( \int_{x_s^e}^{x_e^e} N_i(x)N_j(x) dx \right) \frac{\partial u_j}{\partial t} + N_i(x)\tilde{f} \Big|_{x_s^e}^{x_e^e} - \int_{x_s^e}^{x_e^e} \frac{\partial N_i(x)}{\partial x} f(\hat{u}(x)) dx = 0 \tag{4.10}$$

$$\left( \frac{\Delta x}{2} \int_{-1}^{1} N_i(\xi)N_j(\xi) d\xi \right) \frac{\partial u_j}{\partial t} + N_i(\xi)\tilde{f} \Big|_{-1}^{1} - \int_{-1}^{1} \frac{\partial N_i(\xi)}{\partial \xi} f(\hat{u}(\xi)) d\xi = 0 \tag{4.11}$$

For linear elements, Equation (4.11) can be successively simplified as shown in Equations (4.12) and (4.13). The numerical flux, $\tilde{f}$, is approximated using upwind flux. The upwind flux is calculated based on the eigenvalue $\lambda$. The numerical flux and the eigenvalue are given by the Equations (4.14) and (4.15), respectively.

$$\Delta x \begin{bmatrix} 2/6 & 1/6 \\ 1/6 & 2/6 \end{bmatrix} \frac{\partial}{\partial t} \begin{bmatrix} u_1 \\ u_2 \end{bmatrix} + \begin{bmatrix} -\tilde{f}_1 \\ \tilde{f}_2 \end{bmatrix} - \int_{-1}^{1} f(\hat{u}(\xi)) \, d\xi \begin{bmatrix} -0.5 \\ 0.5 \end{bmatrix} = 0 \tag{4.12}$$

$$\frac{\partial}{\partial t} \begin{bmatrix} u_1 \\ u_2 \end{bmatrix} = \frac{1}{\Delta x} \begin{bmatrix} 4 & -2 \\ -2 & 4 \end{bmatrix} \left( \int_{-1}^{1} f(\hat{u}(\xi)) d\xi \begin{bmatrix} -0.5 \\ 0.5 \end{bmatrix} - \begin{bmatrix} -\tilde{f}_1 \\ \tilde{f}_2 \end{bmatrix} \right) \tag{4.13}$$

$$\tilde{f}(u^-, u^+) = \begin{cases} f(u^-) = 0.5(u^-)^2 & \text{if} \quad \lambda = u > 0 \\ f(u^+) = 0.5(u^+)^2 & \text{if} \quad \lambda = u < 0 \end{cases} \tag{4.14}$$

$$u = 0.5(u^- + u^+) \tag{4.15}$$

For the shock wave case, values of $u_L$ and $u_R$ of 1.0 and 0.5, respectively, are chosen and the numerical solution at $t = 0.6$ is shown in Figure 4.1. The domain is divided into 200 elements of uniform size and a Courant number of 0.1 is used for simulation. The first-order Euler forward time integration and second-order TVD Runge–Kutta (RK2) time integration

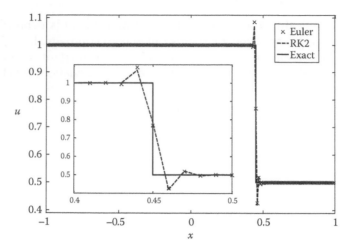

**Figure 4.1** Comparison of numerical and analytical solutions for the shock wave of Burgers' equation.

are evaluated. Oscillations around the shock front are observed in both cases. A numerical technique to get rid of these oscillations with TVD slope limiters will be provided in the next section.

For the rarefaction wave case, values of $u_L$ and $u_R$ of 0.5 and 1.0, respectively, are used in the simulation. The numerical solution at $t = 0.6$ is shown in Figure 4.2 with the same number of elements and Courant number as before. Large oscillations are generated in case of the first-order Euler forward scheme, while the oscillations with second-order TVD Runge–Kutta scheme are negligible.

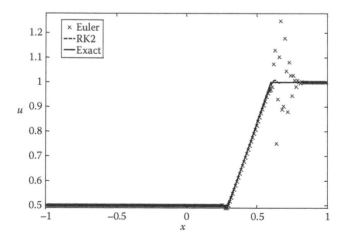

**Figure 4.2** Comparison of numerical and analytical solutions for the rarefaction wave of Burgers' equation.

## 4.2   Total variation diminishing slope limiter

As pointed out in previous examples, unphysical spurious oscillations are observed near shock discontinuities or high gradients with higher-order numerical schemes. The total variation diminishing (TVD) methods are a class of nonlinear methods that provide high-order accuracy and suppress spurious oscillations. Besides the TVD methods, the essentially nonoscillatory (ENO) schemes (Harten et al., 1987) and the weighted essentially nonoscillatory (WENO) schemes (Liu et al., 1994) are widely used to suppress spurious oscillations. The development of TVD methods can be found in literature (Toro, 2009). In this section, the TVD slope limiter is described.

### 4.2.1   TVD methods

The TVD methods have sound theoretical bases only for scalar one-dimensional problems, see for example, Harten (1983), Harten and Lax (1984), and Sweby (1984). However, practical experience and numerical tests demonstrate that the one-dimensional scalar theory works well when extended to nonlinear, multidimensional problems. For a given function $u = u(x)$, the total variation of $u$ is given by Equation (4.16). If $u(x)$ is smooth and differentiable, then Equation (4.16) reduces to Equation (4.17).

$$TV(u) = \lim_{\varepsilon \to 0} \sup \frac{1}{\varepsilon} \int_{-\infty}^{\infty} |u(x+\varepsilon) - u(x)| dx \qquad (4.16)$$

$$TV(u) = \int_{-\infty}^{\infty} \left| \frac{du(x)}{dx} \right| dx \qquad (4.17)$$

Equation (4.17) is still valid even for discontinuous and nondifferential functions, if $du/dx$ is interpreted as distribution derivative. The total variation of $u$ for the discrete case is given by Equation (4.18). A numerical method is called total variation diminishing (TVD) or total variation nonincreasing (TVNI), if the condition given by Equation (4.19) is satisfied. The equation shows progression of solution from a time step $n$ to $n + 1$. If oscillations are produced in a numerical scheme, the total variation will increase in time. The TVD methods suppress these oscillations based on the requirement that the total variation is nonincreasing in time. Thus, TVD methods provide smooth, oscillation-free solutions.

$$TV(u) = \sum_{i=-\infty}^{\infty} |u_{i+1} - u_i| \qquad (4.18)$$

$$TV(u^{n+1}) \le TV(u^n), \quad \forall n \qquad (4.19)$$

## 4.2.2   Formulation of TVD slope limiters

As shown in Section 4.1, the high-order schemes suffer from spurious oscillations in the vicinity of discontinuities and strong gradients. The idea of a slope limiter is to reconstruct the variables with limited slopes using neighboring elements. The data reconstruction method is based on the monotonic upstream scheme for conservation laws (MUSCL) (van Leer, 1979) and the piecewise parabolic method (PPM) (Colella and Woodward, 1984).

The limiting process is applied after the unknowns are calculated in every element in the computational domain. Following the MUSCL scheme, the second-order piecewise linear construction of $U(x)$ within an element is given by Equation (4.20), where $\bar{U}_e$ is the average of the variable $U(x)$ in an element, $U_{el}(x)$ is the limited value within the element, $\bar{x}$ is the midpoint of the element, and $\sigma_e$ is the limited slope in an element $e$. It is obvious that the average value of the variable $U(x)$ would not change after the limiting process, which is an important requirement of conservation laws. The recently computed values are updated with the limited values and the spurious oscillations are suppressed.

$$U_{el}(x) = \bar{U}_e + (x - \bar{x})\sigma_e, \quad x_s^e \le x \le x_e^e \tag{4.20}$$

To calculate the limited slope, the upwind slope $a$, downwind slope $b$, and central slope $(a + b)/2$ must be defined and are given by Equations (4.21) to (4.23), respectively. The MUSCL scheme is TVD if the limited slope is defined by Equation (4.24) (Toro, 2009), where $Cr_e$ is the Courant number inside the element $e$. While the information of Courant number may not be available, several choices of slope limiters that satisfy the TVD condition are reported in the literature (Li, 2006). These limiters include the Godunov method, minmod slope limiter, and monotonized central slope limiter, and are described by Equations (4.25) to (4.27), respectively.

$$a = \frac{\bar{U}_e - \bar{U}_{e-1}}{(\bar{x}_e - \bar{x}_{e-1})} \tag{4.21}$$

$$b = \frac{\bar{U}_{e+1} - \bar{U}_e}{(\bar{x}_{e+1} - \bar{x}_e)} \tag{4.22}$$

$$\frac{a+b}{2} = \frac{\bar{U}_{e+1} - \bar{U}_{e-1}}{(\bar{x}_{e+1} - \bar{x}_{e-1})} \tag{4.23}$$

$$\sigma_e = \left[\frac{sign(a) + sign(b)}{2}\right] \min\left(\beta_1 \left|a\right|, \beta_2 \left|b\right|\right)$$

$$\beta_1 = \frac{2}{1 + Cr_e}$$

$$\beta_2 = \frac{2}{1 - Cr_e}$$

$$(4.24)$$

$$\sigma_e = 0 \tag{4.25}$$

$$\sigma_e = \left[\frac{sign(a) + sign(b)}{2}\right] \min(\left|a\right|, \left|b\right|) \tag{4.26}$$

$$\sigma_e = \left[\frac{sign(a) + sign(b)}{2}\right] \min\left(\frac{\left|a + b\right|}{2}, 2\left|a\right|, 2\left|b\right|\right) \tag{4.27}$$

Numerical results for Burgers' equation with the TVD slope limiters are given in Figures 4.3 and 4.4. The monotonized central (MC) slope limiter is used in the simulation. The slope limiting procedure is applied after every intermediate step of the Runge–Kutta scheme (the slope limiting procedure can also be applied after the Runge–Kutta time integration). The results in Figures 4.3 and 4.4 show that oscillations are eliminated with the

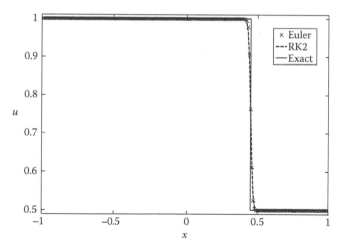

*Figure 4.3* Numerical solutions of the shock wave of Burgers' equation with a slope limiter.

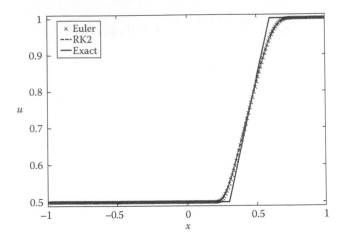

***Figure 4.4*** Numerical solutions of the rarefaction wave of Burgers' equation with a slope limiter.

slope limiter. The results with the Euler forward and TVD Runge–Kutta schemes are similar for the Courant number of 0.1 used in this simulation.

Further investigations show that for the shock wave problem $Cr \leq$ 1.03 is required to achieve stable results with the first-order Euler forward scheme. For the TVD Runge–Kutta scheme, $Cr \leq 1.02$ and $Cr \leq 0.7$ are required for the slope limiter applied after every intermediate step and after the whole time step, respectively. For the rarefaction wave problem, $Cr \leq 1.21$ is required for stability using the first-order Euler forward scheme, whereas $Cr \leq 1.16$ and $Cr \leq 0.48$ are needed with the TVD Runge–Kutta scheme for slope limiter applied after every intermediate time step and after the whole time step, respectively. The TVD slope limiter allows the CFL number to be larger than unity.

Numerical tests show that when the Courant number is relatively large (yet satisfy the stability requirement), the slope limiter applied after every intermediate step in the Runge–Kutta method provides better results than the slope limiter applied after the whole time step. It is recommended that the slope limiter be applied after every intermediate step of the TVD Runge–Kutta scheme to minimize the effect of a large Courant number. The first-order Euler forward scheme provides similar results to the second-order TVD Runge–Kutta scheme, which saves simulation time. In addition, results with the first-order Euler forward scheme are less diffusive as the slope limiter is applied only once, whereas in the two-step second order TVD Runge–Kutta scheme the slope limiter may have to be applied after every intermediate step.

## 4.3 Shallow water flow equations in rectangular channels

The governing equations of open channel flows are conservation of mass and momentum. These two equations are usually referred to as Saint Venant equations. Assuming a rectangular cross-section, mild bed slope, hydrostatic pressure distribution, no lateral inflow or outflow, uniform velocity distribution over the cross-section, constant water density, the longitudinal length scale much larger than the scale for the cross-section, and continuous, differentiable dependent variables, the mass and momentum conservation equations for one-dimensional shallow water flow are given by Equations (4.28) and (4.29), respectively. In these equations, $h$ is the flow depth, $q$ is the flow rate per unit width, and $g$ is the gravitational acceleration. The bed slope, $S_o$, and the friction slope, $S_f$, are given by Equations (4.30) and (4.31), respectively, where $z_b$ is the bed elevation. In Equation (4.31), it is assumed that the friction slope for steady flow is valid under unsteady flow conditions through the Manning equation, where $n$ is the Manning roughness coefficient.

$$\frac{\partial h}{\partial t} + \frac{\partial q}{\partial x} = 0 \tag{4.28}$$

$$\frac{\partial q}{\partial t} + \frac{\partial (q^2/h + gh^2/2)}{\partial x} = gh(S_o - S_f) \tag{4.29}$$

$$S_o = -\frac{\partial z_b}{\partial x} \tag{4.30}$$

$$S_f = \frac{n^2 q |q|}{h^{10/3}} \tag{4.31}$$

The Saint Venant equations can be written in conservation form as shown in Equation (4.32), where $\mathbf{U}$ is the vector of conservative variables, $\mathbf{F}$ is the flux vector, and $\mathbf{S}$ is the source vector. These variables are defined in Equation (4.33). The Jacobian matrix for the Saint Venant equations is given by Equation (4.34). The eigenvalues and independent eigenvectors of the Jacobian matrix are given by Equations (4.35) and (4.36), respectively.

$$\frac{\partial \mathbf{U}}{\partial t} + \frac{\partial \mathbf{F}}{\partial x} = \mathbf{S} \tag{4.32}$$

$$U = \begin{bmatrix} h \\ q \end{bmatrix}$$

$$F = \begin{bmatrix} q \\ gh^2/2 + q^2/h \end{bmatrix} \quad (4.33)$$

$$S = \begin{bmatrix} 0 \\ gh(S_o - S_f) \end{bmatrix}$$

$$A = \frac{\partial F}{\partial U} = \begin{bmatrix} 0 & 1 \\ gh - q^2/h^2 & 2q/h \end{bmatrix} = \begin{bmatrix} 0 & 1 \\ c^2 - u^2 & 2u \end{bmatrix} \quad (4.34)$$

$$\left. \begin{aligned} \lambda_1 &= q/h - \sqrt{gh} = u - c \\ \lambda_2 &= q/h + \sqrt{gh} = u + c \end{aligned} \right\} \quad (4.35)$$

$$\left. \begin{aligned} K_1 &= [1, u - c]^T \\ K_2 &= [1, u + c]^T \end{aligned} \right\} \quad (4.36)$$

The aforementioned eigenvalues are real and distinct for both subcritical and supercritical flows ($h \neq 0$). Thus, the governing equations, given by Equation (4.32), constitute a hyperbolic system. For hyperbolic equations, even with smooth initial and boundary conditions, a discontinuous solution may evolve inside the computational domain. Thus, the numerical model for the shallow water flow equations should be able to capture these discontinuities, and the discontinuous finite element method is an appropriate choice for such problems.

## 4.4  DG method for shallow water flow equations

The computational procedure with the DG method for conservation laws involves dividing the computational domain into a set of elements, deriving the DG formulation for each element, implementing the boundary conditions, if necessary, calculating the numerical flux, time integrating the resulting equations, and finally applying the TVD slope limiter. In this section, the DG formulation for one-dimensional shallow water flow equations in a rectangular channel is provided. Implementation of boundary conditions, evaluation of numerical flux, and application of the TVD slope limiter are also discussed.

### 4.4.1    Shallow water flow equations in rectangular channels

The one-dimensional computational domain ($x = [0, L]$) is divided into $Ne$ elements; let $0 = x_1 < x_2 < \cdots < x_{Ne+1} = L$ be a partition of the domain. If $I_e = [x_s^e, x_e^e]$, $1 \le e \le Ne$, then a discontinuous piecewise finite element space of polynomials $m$ is given by Equation (4.37). Inside an element, the unknowns are approximated by Lagrange interpolation functions as shown in Equation (4.38). The DG formulation for an element is given by Equation (4.39). When the explicit time integration scheme is used, the shallow water flow equations can be solved one by one, Equation (4.39) can be written for each component as given by Equation (4.40).

$$V^m = \{v : v \,|\, I_e \in P^m(I_e),\quad 1 \le e \le Ne\} \tag{4.37}$$

$$\left.\begin{aligned} \mathbf{U} &\simeq \hat{\mathbf{U}} = \sum \mathbf{N}_j(\mathbf{x}) \mathbf{U}_j(\mathbf{x}, t) \\[4pt] \mathbf{F}(\mathbf{U}) &\simeq \hat{\mathbf{F}}(\mathbf{U}) = \mathbf{F}(\hat{\mathbf{U}}) \\[4pt] \mathbf{S}(\mathbf{U}) &\simeq \hat{\mathbf{S}}(\mathbf{U}) = \mathbf{S}(\hat{\mathbf{U}}) \end{aligned}\right\} \tag{4.38}$$

$$\int_{x_s^e}^{x_e^e} \mathbf{N}_i \mathbf{N}_j \, dx \frac{\partial \mathbf{U}_j}{\partial t} + \mathbf{N}_i \tilde{\mathbf{F}}\Big|_{x_s^e}^{x_e^e} - \int_{x_s^e}^{x_e^e} \frac{\partial \mathbf{N}_i}{\partial x} \hat{\mathbf{F}} \, dx = \int_{x_s^e}^{x_e^e} \mathbf{N}_i \hat{\mathbf{S}} \, dx \tag{4.39}$$

$$\int_{x_s^e}^{x_e^e} N_i N_j \, dx \frac{\partial U_j}{\partial t} + N_i \tilde{F}\Big|_{x_s^e}^{x_e^e} - \int_{x_s^e}^{x_e^e} \frac{\partial N_i}{\partial x} \hat{F} \, dx = \int_{x_s^e}^{x_e^e} N_i \hat{S} \, dx \tag{4.40}$$

In Equation (4.40), $U$, $F$, and $S$ are the components of vectors $\mathbf{U}$, $\mathbf{F}$, and $\mathbf{S}$, respectively. For example, $U$ can be $h$ or $q$, and $F$, $S$ can be obtained similarly from Equation (4.33). The DG formulation for the shallow water flow equations can be written for each equation, respectively, as Equations (4.41) and (4.42), where for simplification purposes the flux components are approximated based on Equation (4.43). Using linear elements and transforming the global coordinate into the local coordinate system, Equations (4.41) and (4.42) result in Equations (4.44) and (4.45), respectively. The next step is to evaluate numerical flux $\tilde{F}$ at the element boundaries.

$$\int_{x_s^e}^{x_e^e} N_i N_j \, dx \frac{\partial h_j}{\partial t} + N_i \tilde{f}_1\Big|_{x_s^e}^{x_e^e} - \int_{x_s^e}^{x_e^e} \frac{\partial N_i}{\partial x} \hat{f}_1 \, dx = 0 \tag{4.41}$$

$$\int_{x_s^e}^{x_e^e} N_i N_j \, dx \frac{\partial q_j}{\partial t} + N_i \tilde{f}_2\Big|_{x_s^e}^{x_e^e} - \int_{x_s^e}^{x_e^e} \frac{\partial N_i}{\partial x} \hat{f}_2 \, dx = \int_{x_s^e}^{x_e^e} N_i g \hat{h}(\hat{S}_o - \hat{S}_f) \, dx \tag{4.42}$$

$$\mathbf{F} = \begin{bmatrix} f_1 \\ f_2 \end{bmatrix} = \begin{bmatrix} q \\ gh^2/2 + q^2/h \end{bmatrix} \tag{4.43}$$

$$\Delta x \begin{bmatrix} 2/6 & 1/6 \\ 1/6 & 2/6 \end{bmatrix} \frac{\partial}{\partial t} \begin{bmatrix} h_1 \\ h_2 \end{bmatrix} + \begin{bmatrix} -\tilde{f}_1\big|_{-1} \\ \tilde{f}_1\big|_1 \end{bmatrix} - \begin{bmatrix} -0.5 & -0.5 \\ 0.5 & 0.5 \end{bmatrix} \begin{bmatrix} q_1 \\ q_2 \end{bmatrix} = 0 \tag{4.44}$$

$$\Delta x \begin{bmatrix} 2/6 & 1/6 \\ 1/6 & 2/6 \end{bmatrix} \frac{\partial}{\partial t} \begin{bmatrix} q_1 \\ q_2 \end{bmatrix} + \begin{bmatrix} -\tilde{f}_2\big|_{-1} \\ \tilde{f}_2\big|_1 \end{bmatrix} - \int_{-1}^{1} \frac{\partial N_i(\xi)}{\partial \xi} \hat{f}_2 \, d\xi$$

$$= \frac{\Delta x}{2} \int_{-1}^{1} N_i(\xi) g \hat{h} (\hat{S}_o - \hat{S}_f) \, d\xi \tag{4.45}$$

## 4.4.2 Numerical flux

The HLL flux function is given by Equation (4.46). For the one-dimensional shallow water flow equations, a direct way to calculate the wave speeds ($S_L$ and $S_R$) at the element boundaries is shown in Equation (4.47). Fraccarollo and Toro (1995) suggested estimation of wave speeds based on Equation (4.48), and the definition of the variables are given by Equations (4.49) and (4.50).

$$\mathbf{F}^{HLL} = \begin{cases} \mathbf{F}^- & \text{if} \quad S_L \geq 0 \\ \dfrac{S_R \mathbf{F}^- - S_L \mathbf{F}^+ + S_L S_R (\mathbf{U}^+ - \mathbf{U}^-)}{S_R - S_L} & \text{if} \quad S_L < 0 < S_R \\ \mathbf{F}^+ & \text{if} \quad S_R \leq 0 \end{cases} \tag{4.46}$$

$$\left. \begin{aligned} S_L &= \min\left( u^- - \sqrt{gh^-}, u^+ - \sqrt{gh^+} \right) \\ S_R &= \max\left( u^- + \sqrt{gh^-}, u^+ + \sqrt{gh^+} \right) \end{aligned} \right\} \tag{4.47}$$

$$\left. \begin{aligned} S_L &= \min\left( u^- - \sqrt{gh^-}, u^* - c^* \right) \\ S_R &= \max\left( u^+ + \sqrt{gh^+}, u^* + c^* \right) \end{aligned} \right\} \tag{4.48}$$

$$u^* = \frac{1}{2}(u^- + u^+) + \sqrt{gh^-} - \sqrt{gh^+} \tag{4.49}$$

$$c^* = \frac{1}{2}\left(\sqrt{gh^-} + \sqrt{gh^+}\right) + \frac{1}{4}(u^- - u^+) \tag{4.50}$$

If the wave speeds are defined by Equation (4.51), the resulting flux, given by Equation (4.52), is known as the Lax–Friedrichs flux. The Roe flux function for the one-dimensional shallow water flows in a rectangular channel is given by Equation (4.53). The variables used in the equation are given by Equations (4.54) to (4.58).

$$\left.\begin{array}{c} S_L = -S_{\max} \\[6pt] S_R = S_{\max} \\[6pt] S_{\max} = \max\left(|u^-| + \sqrt{gh^-}, |u^+| + \sqrt{gh^+}\right) \end{array}\right\} \tag{4.51}$$

$$\mathbf{F}^{LF} = \frac{1}{2}(\mathbf{F}^- + \mathbf{F}^+) - S_{\max}(\mathbf{U}^+ - \mathbf{U}^-) \tag{4.52}$$

$$\mathbf{F}^{Roe} = \frac{1}{2}(\mathbf{F}^- + \mathbf{F}^+) - \frac{1}{2}\sum_{i=1}^{2} \tilde{\alpha}_i |\tilde{\lambda}_i| \tilde{\mathbf{K}}_i \tag{4.53}$$

$$\left.\begin{array}{c} \tilde{\alpha}_1 = \frac{1}{2}\left(\Delta h - \frac{\tilde{h}\Delta u}{\tilde{c}}\right) \\[12pt] \tilde{\alpha}_2 = \frac{1}{2}\left(\Delta h + \frac{\tilde{h}\Delta u}{\tilde{c}}\right) \end{array}\right\} \tag{4.54}$$

$$\left.\begin{array}{c} \tilde{\lambda}_1 = \tilde{u} - \tilde{c} \\[6pt] \tilde{\lambda}_2 = \tilde{u} + \tilde{c} \end{array}\right\} \tag{4.55}$$

$$\left.\begin{array}{c} \tilde{\mathbf{K}}_1 = [1, \tilde{u} - \tilde{c}]^T \\[6pt] \tilde{\mathbf{K}}_2 = [1, \tilde{u} + \tilde{c}]^T \end{array}\right\} \tag{4.56}$$

$$\left.\begin{array}{c} \Delta h = h^+ - h^- \\[6pt] \Delta u = u^+ - u^- \end{array}\right\} \tag{4.57}$$

$$\left.\begin{array}{l} \tilde{h} = \sqrt{h^- h^+} \\[2mm] \tilde{u} = \dfrac{\sqrt{h^-}\, u^- + \sqrt{h^+}\, u^+}{\sqrt{h^-} + \sqrt{h^+}} \\[3mm] \tilde{c} = \sqrt{\dfrac{g(h^- + h^+)}{2}} \end{array}\right\} \tag{4.58}$$

### 4.4.3   Dry bed treatment

The shallow water flow equations are strictly hyperbolic for $h > 0$. The treatment for dry bed ($h = 0$) is needed as the numerical flux with approximate Riemann solvers are based on hyperbolic conservation laws. The numerical flux with dry bed should be calculated accurately, and the water depth should be physically correct, that is, $h \geq 0$.

For the dry bed on the right side of a node ($h_L > 0$ and $h_R = 0$), the wave speeds of the HLL Riemann solver are given by Equation (4.59) (Toro, 1990). For the dry bed on the left side of a node ($h_L = 0$ and $h_R > 0$), the wave speeds are given by Equation (4.60). For Roe's solver, the wave speeds for the dry bed on the right side or left side of a node are given, respectively, by Equations (4.61) and (4.62). For the previous equations, $h_L = h^-$ and $h_R = h^+$.

$$\left.\begin{array}{l} S_L = u_L - \sqrt{gh_L} \\[2mm] S_R = u_L + 2\sqrt{gh_L} \end{array}\right\} \tag{4.59}$$

$$\left.\begin{array}{l} S_L = u_R - 2\sqrt{gh_R} \\[2mm] S_R = u_R + \sqrt{gh_R} \end{array}\right\} \tag{4.60}$$

$$\left.\begin{array}{l} \tilde{\lambda}_1 = u_L - \sqrt{gh_L} \\[2mm] \tilde{\lambda}_2 = u_L + 2\sqrt{gh_L} \end{array}\right\} \tag{4.61}$$

$$\left.\begin{array}{l} \tilde{\lambda}_1 = u_R - 2\sqrt{gh_R} \\[2mm] \tilde{\lambda}_2 = u_R + \sqrt{gh_R} \end{array}\right\} \tag{4.62}$$

To handle the dry bed problem, a small depth $\varepsilon$ (e.g., $\varepsilon = 10^{-16}$) can be used to check the wet/dry front (Sanders, 2001), and the water depth at the dry nodes is set to zero. If the water depth at one side of the element face is larger than ε and the other side is less than or equal to ε, the numerical flux across this element face is computed according to the dry bed location. If the water depth at both sides are less than or equal to ε, the numerical flux is set to zero. In the event that the computed water depth is less than or equal to ε, the velocity at that node is set to zero. If the computed water depth is less than zero, then both the depth and velocity at that node are set to zero.

In the second method, a sufficiently small depth $h_{dry}$ (e.g., $10^{-16}$) and zero velocity can be defined at the dry nodes (Ying et al., 2003, 2004). At the element boundary, if the water depth at one side is greater than $h_{dry}$ and the other side is less than or equal to $h_{dry}$, then the numerical flux is calculated according to the dry bed location. If the water depth on both sides of the element boundary is less than or equal to $h_{dry}$, then the numerical flux which is calculated with the HLL or Roe flux are set to zero. After every time step, the water depth at every node is checked. If the water depth at a node is less than $h_{dry}$, the water depth is set to $h = h_{dry}$ and the velocity is set to zero.

For horizontal beds, these two methods will give the same results. In the case of a channel with a bed variation, even a predefined sufficiently small depth $h_{dry}$ may generate an unphysical flow. So, the first method is recommended for dry bed problems with a bed variation. However, in the first method, since the depth is zero at the dry nodes, special care should be taken with terms divided by $h$, for example, $q^2/h$ and the friction term.

### 4.4.4    Initial and boundary conditions

Initial and boundary conditions are essential for the solution of shallow water flow equations. The flow conditions at the start of the computation are referred to as initial conditions. The flow conditions at the boundaries of the computational domain are called boundary conditions. The requirements and implementation of initial and boundary conditions are discussed in this section. More discussion and advanced topics about Riemann invariants and characteristics can be found in the literature (e.g., Stoker, 1957; Cunge et al., 1980).

The characteristic directions of subcritical flow and supercritical flow at the boundaries of the 1D flow domain are shown in Figure 4.5 and Figure 4.6, respectively. The initial conditions can be considered as special cases of boundary conditions. For any point inside the domain, the solution of the water depth $h$ and flow rate $q$ requires the flow information from characteristics, $\psi_1$ and $\psi_2$. The characteristic directions are mathematically represented by the eigenvalues $(\psi_{1,2} = \lambda_{1,2} = u \pm c)$. While at the boundaries of a flow domain, information of the characteristics

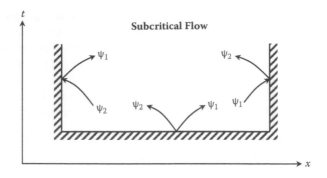

**Figure 4.5** Characteristics at the boundaries for subcritical flow.

may be missing, so boundary conditions are required for the solution of the governing equations. The number of boundary conditions required depends on the number of characteristics entering the domain from a given boundary. The boundary conditions should be independent of the governing equations and the Riemann invariants, and the boundary conditions should be independent of each other if two conditions are required (Cunge et al., 1980).

Following the aforementioned rules, for the solution of the 1D shallow water flow equations, two initial conditions are required, such as the water depth $h$ and the flow rate $q$, or the water depth $h$ and the flow velocity $u$. The inflow and outflow boundary conditions needed for subcritical, critical, and supercritical flows are listed in Table 4.1. At the inflow boundary, one condition is needed for subcritical flow and two are required for supercritical flow. At the outflow boundary, one boundary condition is

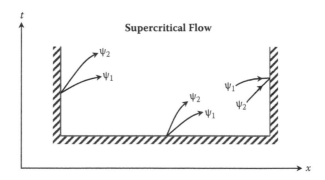

**Figure 4.6** Characteristics at the boundaries for a supercritical flow.

*Table 4.1* Required Boundary Conditions for 1D Shallow Water Flow Equations

| Flow Regime | Inflow Boundary | Outflow Boundary | Initial Condition |
|---|---|---|---|
| Subcritical | 1 | 1 | 2 |
| Supercritical | 2 | 0 | 2 |
| Critical | 1 | 0 | 2 |

required for a subcritical flow and none for a supercritical flow. For critical flow, one boundary condition is needed at the inflow boundary, and the other is determined from the Froude number relationship.

## 4.5   Numerical tests

In this section, numerical solutions for shallow water flow equations with the discontinuous Galerkin method are presented. Numerical tests include idealized dam break, hydraulic jump, flow over a bump, and flow over an irregular bed. The numerical solutions are compared with analytical solutions for idealized cases and laboratory data (if available).

### 4.5.1   Idealized dam break in a rectangular channel

The discontinuous Galerkin method is used to model idealized dam-break flow in a rectangular channel with a horizontal bed. The exact solution for these idealized dam-break problems can be found in Henderson (1966). The channel is 1000 m in length, with a dam located at 500 m. The water depth upstream of the dam is 10 m, and the water depth downstream is 2 m, and 0 for wet bed and dry bed cases, respectively. The dam is removed instantaneously and the water flow afterward is simulated. The computational domain is discretized into 400 elements for both cases. The water surface and flow rate at 20 s after the dam break are shown in Figures 4.7 to 4.10 for the wet bed and dry bed cases. Numerical results show similar performance with the HLL, LF, and Roe fluxes, so the results with HLL flux are shown. For a dam break with a wet bed downstream, the simulated results are in excellent agreement with the exact solution. For the dry bed case, the results are in good agreement with the exact solution except at the front of the wave.

### 4.5.2   Dam break in a rectangular flume with bed friction

In this test, the numerical scheme is used to simulate the dam-break flow in a rectangular, horizontal flume, where the measured water surface

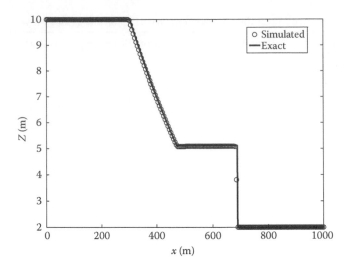

**Figure 4.7** Comparison of water surface profiles for a dam break over a wet bed in a rectangular channel.

profiles are available (Schoklitsch, 1917). The flume is 0.096 m in width, 0.08 m in height, and 20 m in length, with a dam located at 10 m. For the flume, made of smooth wood, the Manning coefficient of 0.009 s/m$^{1/3}$ is used (Khan, 2000). The water surface elevation is 0.074 m upstream of the dam with a dry bed downstream. The removal of the dam is assumed

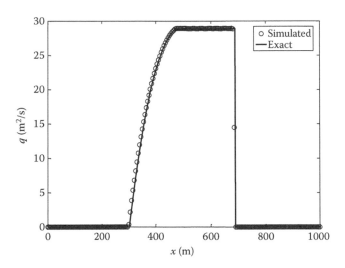

**Figure 4.8** Comparison of flow rates for a dam break over a wet bed in a rectangular channel.

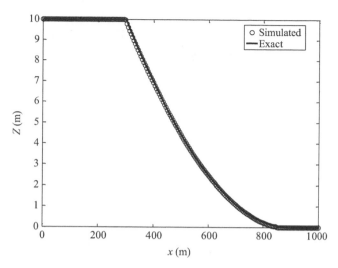

**Figure 4.9** Comparison of water surface profiles for a dam break over a dry bed in a rectangular channel.

instantaneous and the flow afterward is simulated. The simulated and measured water surface profiles at 3.75 s and 9.40 s after the dam removal are shown in Figure 4.11. The simulated water surface profiles are in excellent agreement with the measured data, showing the scheme is capable to model a dam-break flow over an initially dry bed with friction.

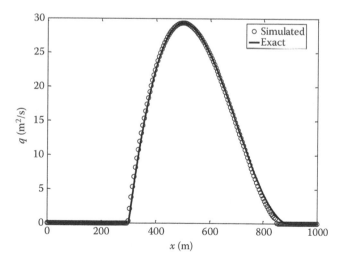

**Figure 4.10** Comparison of flow rates for a dam break over a dry bed in a rectangular channel.

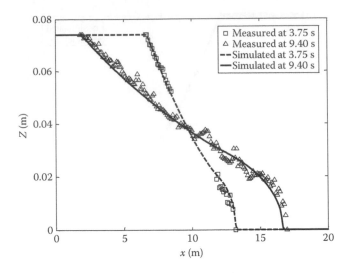

**Figure 4.11** Simulated and measured water surface profiles after a dam break with bed friction.

### 4.5.3   Hydraulic jump

A numerical simulation of a hydraulic jump is performed in this test. The numerical simulations of water surface profiles are compared with the measured data collected by Gharangik and Chaudhry (1991). The channel is 14 m long and 0.46 m wide with a horizontal bed. The Manning roughness coefficient is taken to be 0.008 s/m$^{1/3}$. The initial water depth is 0.031 m with a zero flow rate. At the upstream inflow boundary, the water depth of 0.031 m and discharge of 0.118 m$^2$/s are specified. At the downstream boundary, the water depth is increased from 0.031 m to 0.265 m in 50 s and kept constant at 0.265 m. The domain is discretized with 100 elements and the Courant number of 0.1 is used in this simulation. The steady-state solutions of the water surface and flow rate are presented in Figure 4.12 and Figure 4.13, respectively. The numerical scheme is capable of capturing shocks and preserving the mass flux. The water surface elevation results in the LF flux function having oscillations at the jump location. The numerical schemes with HLL and Roe flux give oscillation-free results.

### 4.5.4   Flow over a bump

Numerical simulations of flow over a bump with different flow regimes are performed. The frictionless rectangular channel is 1 m wide and 25 m long with the bed elevation defined by Equation (4.63) (Goutal and Maurel, 1997). For the subcritical flow over the bump, the initial water

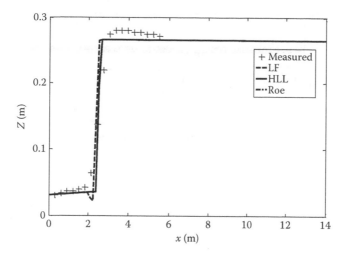

**Figure 4.12** Comparison of water surface profiles for a hydraulic jump.

surface is 0.5 m with still water. The flow rate at the upstream boundary is 0.18 m²/s and water depth at the outflow is set to 0.5 m. Thus, a subcritical flow condition exists throughout the channel.

$$z_b = \begin{cases} 0.2 - 0.05(x - 10)^2, & 8 \leq x \leq 12 \\ 0, & \text{otherwise} \end{cases} \qquad (4.63)$$

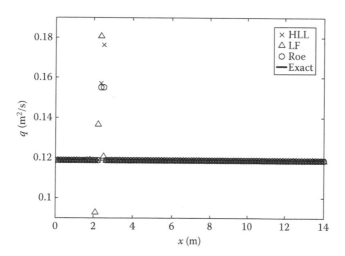

**Figure 4.13** Mass conservation in a hydraulic jump.

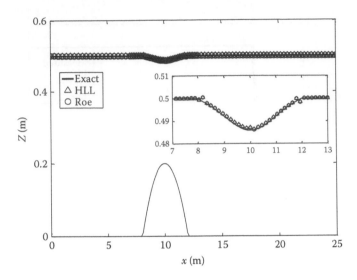

**Figure 4.14** Comparison of water surface profiles for a subcritical flow over the bump with the water-depth-based slope limiter.

The numerical results with the slope limiter applied to the water depth and flow rate are shown in Figures 4.14 and 4.15, whereas Figures 4.16 and 4.17 show numerical results with the slope limiter applied to the water surface and flow rate. With the water-depth-based slope limiter, oscillations are observed in the flow rate, and flow rate is conserved with the slope limiter

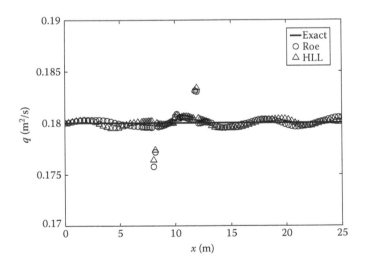

**Figure 4.15** Comparison of flow rates for a subcritical flow over the bump with the water-depth-based slope limiter.

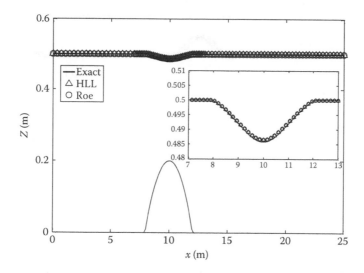

*Figure 4.16* Comparison of water surface profiles for a subcritical flow over the bump with the water-surface-based slope limiter.

applied to the water surface. The computed water depth and flow rate with the water-surface-based slope limiter are in good agreement with the exact solution. The HLL flux and Roe flux provide similar results in this case.

For the supercritical flow over the bump, the initial water surface is 2.0 m with still water. At the inflow boundary, the flow rate of 25.0567 m²/s and

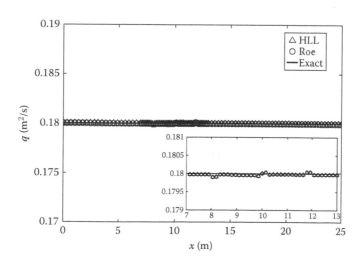

*Figure 4.17* Comparison of flow rates for a subcritical flow over the bump with the water-surface-based slope limiter.

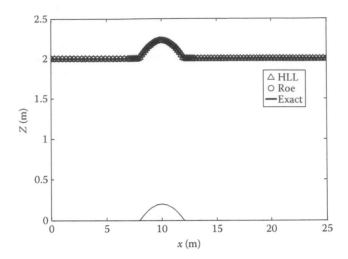

***Figure 4.18*** Comparison of water surface profiles for a supercritical flow over the bump with the water-surface-based slope limiter.

water depth of 2.0 m are specified, resulting in supercritical flow through-out the channel. The numerical results are shown in Figures 4.18 and 4.19 with the water-surface-based slope limiter. The simulated results are in excellent agreement with the exact solutions. As before, the performance of HLL flux and Roe flux are similar.

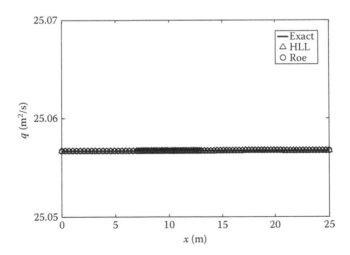

***Figure 4.19*** Comparison of flow rates for a supercritical flow over the bump with the water-surface-based slope limiter.

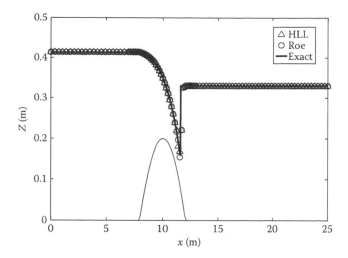

***Figure 4.20*** Comparison of water surface profiles for a transcritical flow over the bump with the water-surface-based slope limiter.

Transcritical flow over the bump is simulated using initial water surface elevation of 0.33 m with still water. The flow rate at the upstream boundary is 0.18 m²/s and the water depth at the outflow boundary is set to 0.33 m. The flow changes from a subcritical flow to a supercritical flow and back to a subcritical flow through a hydraulic jump. The numerical results are shown in Figures 4.20 and 4.21 with the water-surface-based

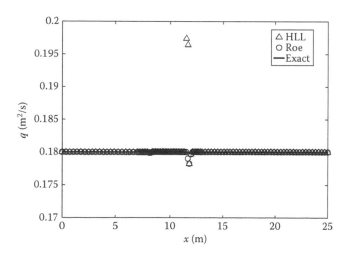

***Figure 4.21*** Comparison of flow rates for a transcritical flow over the bump with the water-surface-based slope limiter.

*Table 4.2* Bed Elevation Variation with Distance

| $x$ (m) | 0 | 50 | 100 | 150 | 200 | 250 | 300 | 350 | 400 | 425 |
|---|---|---|---|---|---|---|---|---|---|---|
| $z_b$ (m) | 0 | 0 | 2.5 | 5 | 5 | 3 | 5 | 5 | 7.5 | 8 |
| $x$ (m) | 435 | 450 | 470 | 475 | 500 | 505 | 530 | 550 | 565 | 575 |
| $z_b$ (m) | 9 | 9 | 9 | 9.1 | 9 | 9 | 6 | 5.5 | 5.5 | 5 |
| $x$ (m) | 600 | 650 | 700 | 750 | 800 | 820 | 900 | 950 | 1000 | 1500 |
| $z_b$ (m) | 4 | 3 | 3 | 2.3 | 2 | 1.2 | 0.4 | 0 | 0 | 0 |

slope limiter. The location of the hydraulic jump is predicted accurately and the flow rate is well conserved except at the jump. The Roe flux performs better than the HLL flux within the jump. Overall, the numerical results are satisfactory.

### 4.5.5    Flow over irregular bed

In this test, transcritical flow in a channel with an irregular, frictionless bed is simulated. The bed elevation is given in Table 4.2 (Goutal and Maurel, 1997). The initial conditions consist of a water surface elevation of 16 m with zero velocity. The flow rate at the inflow section is set to 50 m²/s and the water surface at the downstream end is maintained at 16 m. The flow within the computational domain changes from subcritical flow to supercritical flow and back to subcritical flow through a hydraulic jump. Numerical results are shown in Figures 4.22 and 4.23. The flow rate is better conserved with the Roe flux than the HLL flux within the jump.

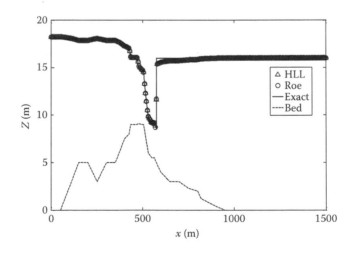

*Figure 4.22* Comparison of water surface profiles for a transcritical flow over an irregular bed.

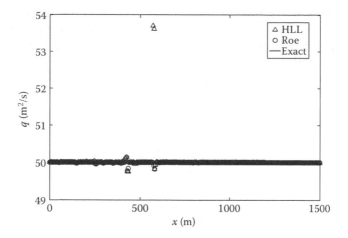

**Figure 4.23** Comparison of flow rates for a transcritical flow over an irregular bed.

## 4.5.6    *Wetting and drying in a parabolic bowl*

Thacker (1981) developed analytical solutions for the one-dimensional flow in a frictionless, parabolic bowl with a moving shoreline. These analytical solutions are useful for numerical tests including wetting and drying. The parabolic bed profile of a rectangular channel with a unit width is given by Equation (4.64), where the constants $h_o$ and $a$ are depicted in Figure 4.24.

$$z_b(x) = h_o \left( \frac{x^2}{a^2} - 1 \right) \tag{4.64}$$

The analytic solutions for the water surface and velocity in the region where the water depth is nonzero are given by Equations (4.65) and (4.66),

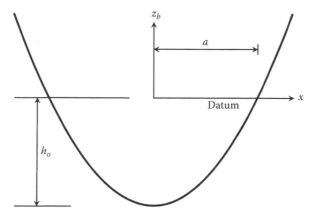

**Figure 4.24** Bed profile of the parabolic bowl.

respectively, where $B$ is a constant and $\omega$ is given by Equation (4.67). The location of the moving shoreline at any time is given by Equation (4.68). In this example, $h_o = 10$ m, $a = 600$ m, and $B = 5$ m/s, which give an oscillation period of $T = 269$ s.

$$Z(x,t) = \frac{-B^2 \cos(2\omega t) - B^2 - 4B\omega \cos(\omega t)x}{4g} \tag{4.65}$$

$$u(x,t) = B \sin(\omega t) \tag{4.66}$$

$$\omega = \frac{2\pi}{T} = \frac{\sqrt{2gh_o}}{a} \tag{4.67}$$

$$x = -\frac{a^2 \omega B}{2gh_o} \cos(\omega t) \pm a \tag{4.68}$$

The computational domain spans over $x = [-1000, 1000]$ and is discretized using 2000 elements. The initial conditions are obtained from Equation (4.65) with zero velocity. The initial water surface is shown in Figure 4.25. The HLL flux function is used with the estimated wave speeds based on Equation (4.69). Simulated and exact solutions of the water surface and the flow rate at different times are compared in Figures 4.26 to 4.33. The numerical results are in good agreement with exact solutions.

$$\left.\begin{aligned} S_L &= \min\left(u^- - \sqrt{gh^-}, u^+ - \sqrt{gh^+}\right) \\ S_R &= \max\left(u^- + \sqrt{gh^-}, u^+ + \sqrt{gh^+}\right) \end{aligned}\right\} \tag{4.69}$$

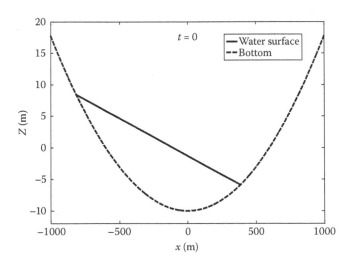

*Figure 4.25* Initial water surface profile in the parabolic bowl.

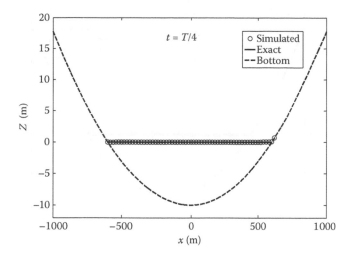

**Figure 4.26** Simulated and exact water surface profiles in the parabolic bowl at $t = T/4$.

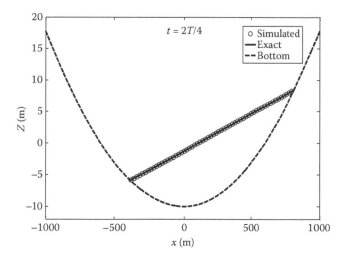

**Figure 4.27** Simulated and exact water surface profiles in the parabolic bowl at $t = 2T/4$.

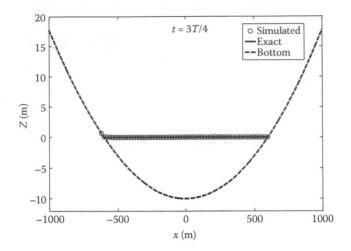

*Figure 4.28* Simulated and exact water surface profiles in the parabolic bowl at $t = 3T/4$.

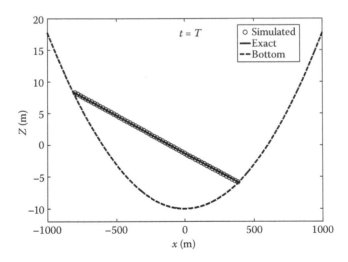

*Figure 4.29* Simulated and exact water surface profiles in the parabolic bowl at $t = T$.

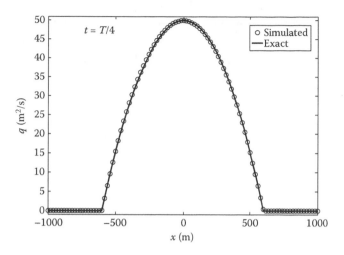

**Figure 4.30** Simulated and exact flow rates in the parabolic bowl at $t = T/4$.

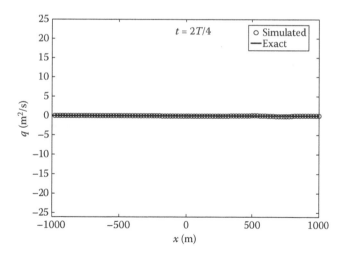

**Figure 4.31** Simulated and exact flow rates in the parabolic bowl at $t = 2T/4$.

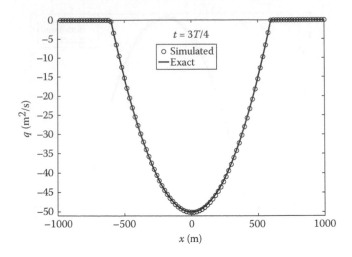

**Figure 4.32** Simulated and exact flow rates in the parabolic bowl at $t = 3T/4$.

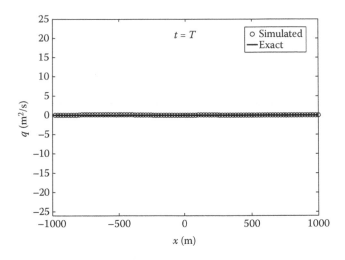

**Figure 4.33** Simulated and exact flow rates in the parabolic bowl at $t = T$.

# References

Colella, P., and Woodward, P. R. (1984). The piecewise parabolic method (PPM) for gas-dynamical simulations. *Journal of Computational Physics*, 54(1), 174–201.

Cunge, J. A., Holly, F. M. Jr., and Verwey, A. (1980). *Practical Aspects of Computational River Hydraulics*. Pitman, London.

Fraccarollo, L., and Toro, E. F. (1995). Experimental and numerical assessment of the shallow water model for two-dimensional dam-break type problems. *Journal of Hydraulic Research*, 33(6), 843–864.

Gharangik, A. M., and Chaudhry, M. H. (1991). Numerical simulation of hydraulic jump. *Journal of Hydraulic Engineering*, 117(9), 1195–1211.

Goutal, N., and Maurel, F. (1997). Proceedings of the 2nd workshop on dam-break wave simulation. HE-43/97/016/B, Direction des Études et Recherches, EDF.

Harten, A. (1983). High resolution schemes for hyperbolic conservation laws. *Journal of Computational Physics*, 49(3), 357–393.

Harten, A., Engquist, B., Osher, S., and Chakravarthy, S. R. (1987). Uniformly high order accurate essentially non-oscillatory schemes, III. *Journal of Computational Physics*, 71(2), 231–303.

Harten, A., and Lax, P. D. (1984). On a class of high resolution total-variation-stable finite-difference schemes. *SIAM Journal on Numerical Analysis*, 21(1), 1–23.

Henderson, F. M. (1966). *Open Channel Flow*. McGraw-Hill, New York.

Li, B. Q. (2006). *Discontinuous Finite Elements in Fluid Dynamics and Heat Transfer*. Springer-Verlag, London.

Liu, X. D., Osher, S., and Chan, T. (1994). Weighted essentially non-oscillatory schemes. *Journal of Computational Physics*, 115(1), 200–212.

Sanders, B. F. (2001). High-resolution and non-oscillatory solution of the St. Venant equations in non-rectangular and non-prismatic channels. *Journal of Hydraulic Research*, 39(3), 321–330.

Schoklitsch, A. (1917). Ueber dammbruchwellen. *Sitzungsberichte der Kaiserlichen Akademie Wissenschaften, Viennal*, 126, 1489–1514.

Stoker, J. J. (1957). *Water Waves*. Interscience, New York.

Sweby, P. K. (1984). High resolution schemes using flux limiters for hyperbolic conservation laws. *SIAM Journal on Numerical Analysis*, 21(5), 995–1011.

Thacker, W. C. (1981). Some exact solutions to the nonlinear shallow-water wave equations. *Journal of Fluid Mechanics*, 107, 499–508.

Toro, E. F. (1990). The dry-bed problem in shallow-water flows. College of Aeronautics, Report No. 9007, Cranfield Institute of Technology, Cranfield, U.K.

Toro, E. F. (2009). *Riemann Solvers and Numerical Methods for Fluid Dynamics*, 3rd ed. Springer-Verlag, Berlin, Heidelberg.

van Leer, B. (1979). Towards the ultimate conservative difference scheme, V: A second-order sequel to Godunov's method. *Journal of Computational Physics*, 32(1), 101–136.

Ying, X., Khan, A. A., and Wang, S. S. Y. (2003). An upwind method for one-dimensional dam break flows. Proceedings of XXX IAHR Congress, Thessaloniki, Greece, August 24–29.

Ying, X., Khan, A. A., and Wang, S. S. Y. (2004). Upwind conservative scheme for the Saint Venant equations. *Journal of Hydraulic Engineering*, 130(10), 977–987.

# chapter five

## One-dimensional shallow water flow in nonrectangular channels

In this chapter, the discontinuous Galerkin (DG) method is applied to model one-dimensional shallow water flow in nonrectangular, nonprismatic channels. The governing equations for one-dimensional shallow water flows in natural channels are first discussed. Then, the discontinuous Galerkin method is applied to solve these equations.

### 5.1  General form of the Saint Venant equations

The one-dimensional model for shallow water flows in nonrectangular channels is based on the assumptions similar to those described for rectangular channels, except that the cross-sections are of arbitrary shape. The governing equations for one-dimensional shallow water flows in nonrectangular, nonprismatic channels are referred to as the Saint Venant equations and include the conservation of mass and momentum (Cunge et al., 1980). The equations for the mass and momentum conservations are given by Equations (5.1) and (5.2), respectively. The hydrostatic pressure force ($I_1$), wall pressure force ($I_2$), bed slope ($S_o$), and friction slope ($S_f$) are defined by Equations (5.3) and (5.4).

$$\frac{\partial A}{\partial t} + \frac{\partial Q}{\partial x} = 0 \tag{5.1}$$

$$\frac{\partial Q}{\partial t} + \frac{\partial (Q^2/A + gI_1)}{\partial x} = gI_2 + gA(S_o - S_f) \tag{5.2}$$

$$\left. \begin{aligned} I_1 &= \int_0^{h(x,t)} (h-y)b(x,y)\,dy \\ I_2 &= \int_0^{h(x,t)} (h-y)\frac{\partial b(x,y)}{\partial x}\,dy \end{aligned} \right\} \tag{5.3}$$

$$\left. \begin{aligned} S_o &= -\frac{\partial z_b}{\partial x} \\ S_f &= \frac{n^2 Q|Q|}{R^{4/3}A^2} \end{aligned} \right\} \tag{5.4}$$

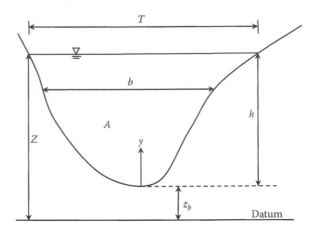

***Figure 5.1*** Cross-section geometry of a natural river.

A sketch of a typical channel cross-section is shown in Figure 5.1. In the preceding equations, $Q$ is the flow rate, $A$ is the cross-section area, $h$ is the water depth, $z_b$ is the channel bed elevation from the datum, $Z$ is the water surface elevation ($Z = z_b + h$), $n$ is the Manning roughness coefficient, $g$ is the gravitational acceleration, $b$ is the channel width at any depth, $R$ is the hydraulic radius, and $T$ is the width of the channel at the water surface.

The governing equations can also be written in the conservation form as given by Equation (5.5), where $\mathbf{U}$ is the vector of conserved variables, $\mathbf{F(U)}$ is the flux vector, and $\mathbf{S(U)}$ is the vector of source terms as defined by Equation (5.6). The Jacobian matrix is given by Equation (5.7). Garcia-Navarro and Vazquez-Cendon (2000) showed that, in the Jacobian matrix, the hydrostatic pressure force term can be approximated by Equation (5.8). The wave celerity, $c$, is also defined in the same equation. The eigenvalues and the corresponding eigenvectors of the Jacobian matrix are given by Equations (5.9) and (5.10), respectively.

$$\frac{\partial \mathbf{U}}{\partial t} + \frac{\partial \mathbf{F(U)}}{\partial x} = \mathbf{S(U)} \tag{5.5}$$

$$\left.\begin{aligned} \mathbf{U} &= \begin{bmatrix} A \\ Q \end{bmatrix} \\[1em] \mathbf{F} &= \begin{bmatrix} Q \\ Q^2/A + gI_1 \end{bmatrix} \\[1em] \mathbf{S} &= \begin{bmatrix} 0 \\ gI_2 + gA(S_o - S_f) \end{bmatrix} \end{aligned}\right\} \tag{5.6}$$

$$J = \frac{\partial F}{\partial U} = \begin{bmatrix} 0 & 1 \\ g\frac{A}{T} - \left(\frac{Q}{A}\right)^2 & \frac{2Q}{A} \end{bmatrix} = \begin{bmatrix} 0 & 1 \\ c^2 - u^2 & 2u \end{bmatrix} \qquad (5.7)$$

$$\left. \begin{aligned} \frac{\partial I_1}{\partial A} &= \frac{A}{T} \\[2mm] c &= \sqrt{g\frac{A}{T}} \end{aligned} \right\} \qquad (5.8)$$

$$\left. \begin{aligned} \lambda_1 &= u - c \\ \lambda_2 &= u + c \end{aligned} \right\} \qquad (5.9)$$

$$\left. \begin{aligned} K_1 &= [1, u - c]^T \\ K_2 &= [1, u + c]^T \end{aligned} \right\} \qquad (5.10)$$

The flow in a channel with a rectangular section, discussed in the previous chapter, is a particular case of natural river cross-section (for a rectangular channel $b(x, y) \equiv T$). Due to the difficulty in computing the general hydrostatic and wall pressure force terms in nonrectangular and nonprismatic channels, the two terms can be simplified using Leibnitz's rule as shown in Equation (5.11). The resulting momentum equation is given by Equation (5.12).

$$\frac{\partial I_1}{\partial x} - I_2 = A\frac{\partial h}{\partial x} \qquad (5.11)$$

$$\frac{\partial Q}{\partial t} + \frac{\partial(Q^2/A)}{\partial x} = -gA\frac{\partial Z}{\partial x} - gAS_f \qquad (5.12)$$

In Equation (5.12), the hydrostatic pressure and wall pressure terms are combined. With the simplified momentum equation, given by Equation (5.12), the well-balanced property is automatically preserved for the wet bed case, that is, initially still water with a horizontal water surface will remain still irrespective of the bed topography. Thus, this formulation circumvents the numerical generated flow due to inadequate treatment of the bed slope term (Ying et al., 2004).

The continuity equation, Equation (5.1), and the simplified momentum equation, Equation (5.12), can also be written in the conservation form as given by Equation (5.13). The vectors of conserved variables, flux, and source are defined in Equation (5.14). Following the analysis method described earlier, the Jacobian matrix of Equation (5.13) is given by Equation (5.15). The eigenvalues of the Jacobian matrix are given by Equation (5.16). The eigenvalues are the same, which violates the hyperbolic property of the original Saint Venant equations given by Equation (5.5). The flow properties are governed by the eigenvalues given in Equation (5.9). For this reason, in the evaluation of the numerical flux, the eigenvalues and eigenvectors given in Equations (5.9) and (5.10), respectively, are used.

$$\frac{\partial \mathbf{U}}{\partial t} + \frac{\partial \mathbf{F}(\mathbf{U})}{\partial x} = \mathbf{S}(\mathbf{U}) \tag{5.13}$$

$$\left. \begin{aligned} \mathbf{U} &= \begin{bmatrix} A \\ Q \end{bmatrix} \\[2em] \mathbf{F} &= \begin{bmatrix} Q \\ Q^2/A \end{bmatrix} \\[2em] \mathbf{S} &= \begin{bmatrix} 0 \\ -gA\dfrac{\partial Z}{\partial x} - gAS_f \end{bmatrix} \end{aligned} \right\} \tag{5.14}$$

$$\mathbf{J} = \frac{\partial \mathbf{F}}{\partial \mathbf{U}} = \begin{bmatrix} 0 & 1 \\ -\left(\dfrac{Q}{A}\right)^2 & \dfrac{2Q}{A} \end{bmatrix} = \begin{bmatrix} 0 & 1 \\ -u^2 & 2u \end{bmatrix} \tag{5.15}$$

$$\lambda_1 = \lambda_2 = u \tag{5.16}$$

## 5.2  *Discontinuous Galerkin method for general Saint Venant equations*

In this section, the discontinuous Galerkin (DG) formulation for Saint Venant equations is presented. The one-dimensional domain ($x = [0, L]$) is divided into $Ne$ elements such that $0 = x_1 < x_2 < \cdots < x_{Ne+1} = L$. A typical element is given by $I_e = [x_s^e, x_e^e]$, $1 \le e \le Ne$. Inside an element, unknowns are approximated by Lagrange interpolation functions as given by Equation (5.17). The governing equation is multiplied by the weight or test function, and the resulting equations are integrated over an element. The flux term is integrated by parts, and the final equation is shown in Equation (5.18) and can be rewritten in a compact form as Equation (5.19).

$$
\left.\begin{aligned}
\mathbf{U} \simeq \hat{\mathbf{U}} = \sum \mathbf{N}_j(\mathbf{x})\mathbf{U}_j(\mathbf{x},t) \\[4pt]
\mathbf{F}(\mathbf{U}) \simeq \hat{\mathbf{F}} = \mathbf{F}(\hat{\mathbf{U}}) \\[4pt]
\mathbf{S}(\mathbf{U}) \simeq \hat{\mathbf{S}} = \mathbf{S}(\hat{\mathbf{U}})
\end{aligned}\right\} \tag{5.17}
$$

$$
\int_{x_s^e}^{x_e^e} \mathbf{N}_i \mathbf{N}_j \, dx \frac{\partial \mathbf{U}_j}{\partial t} + \mathbf{N}_i \tilde{\mathbf{F}} \Big|_{x_s^e}^{x_e^e} - \int_{x_s^e}^{x_e^e} \frac{\partial \mathbf{N}_i}{\partial x} \hat{\mathbf{F}} \, dx = \int_{x_s^e}^{x_e^e} \mathbf{N}_i \hat{\mathbf{S}} \, dx \tag{5.18}
$$

$$
\mathbf{M}\frac{\partial \mathbf{U}}{\partial t} = \mathbf{R} \quad \text{or} \quad \frac{\partial \mathbf{U}}{\partial t} = \mathbf{M}^{-1}\mathbf{R} = \mathbf{L} \tag{5.19}
$$

The solution of conserved variable $\mathbf{U}$ can be obtained with an appropriate time integration scheme such as a total variation diminishing (TVD) Runge–Kutta scheme. The TVD slope limiter is also needed to prevent unphysical oscillations with higher-order schemes. For the numerical flux, the approximate Riemann solver, such as HLL (Harten-Lax-Van Leer) or Roe method, can be used. In the Saint Venant equations, the conserved variables vector and the flux vector are, respectively, given by $\mathbf{U} = [A, Q]^T$ and $\mathbf{F} = [Q, Q^2/A]^T$. The HLL flux is given by Equation (5.20). The left and right wave speeds ($S_L$ and $S_R$, respectively) are given by Equation (5.21) as suggested by Fraccarollo and Toro (1995), where $u^*$ and $c^*$ are defined in Equation (5.22) and Equation (5.23), respectively.

$$
\mathbf{F}^{HLL} = \begin{cases}
\mathbf{F}_L & \text{if} \quad S_L \geq 0 \\[6pt]
\dfrac{S_R \mathbf{F}_L - S_L \mathbf{F}_R + S_L S_R (\mathbf{U}_R - \mathbf{U}_L)}{S_R - S_L} & \text{if} \quad S_L < 0 < S_R \\[10pt]
\mathbf{F}_R & \text{if} \quad S_R \leq 0
\end{cases} \tag{5.20}
$$

$$
\left.\begin{aligned}
S_L = \min\left(u^- - \sqrt{g(A/T)^-}, u^* - c^*\right) \\[6pt]
S_R = \max\left(u^+ + \sqrt{g(A/T)^+}, u^* + c^*\right)
\end{aligned}\right\} \tag{5.21}
$$

$$
u^* = \frac{1}{2}(u^- + u^+) + \sqrt{g(A/T)^-} - \sqrt{g(A/T)^+} \tag{5.22}
$$

$$
c^* = \frac{1}{2}\left(\sqrt{g(A/T)^-} + \sqrt{g(A/T)^+}\right) + \frac{1}{4}(u^- - u^+) \tag{5.23}
$$

Following Garcia-Navarro and Vazquez-Cendon (2000), the Roe's flux for the Saint Venant equations for nonrectangular channels is given by Equation (5.24), where the variables in the equation are defined by Equations (5.25) to (5.29). As pointed out before, the shallow water flow is governed by the eigenvalues of the Saint Venant equations given by Equations (5.5) and (5.6), and represent physical characteristics of the flow. The choice of the equation modeled should not affect the eigenvalues. Thus, the wave speeds in Equations (5.21) and (5.26) are approximated with the eigenvalues given by Equation (5.9).

$$\mathbf{F}^{Roe} = \frac{1}{2}(\mathbf{F}^- + \mathbf{F}^+) - \frac{1}{2}\sum_{i=1}^{2} \tilde{\alpha}_i \, |\tilde{\lambda}_i| \, \tilde{\mathbf{K}}_i \qquad (5.24)$$

$$\left. \begin{aligned} \tilde{\alpha}_1 &= \frac{(\tilde{c} + \tilde{u})\Delta A - \Delta Q}{2\tilde{c}} \\[2mm] \tilde{\alpha}_2 &= \frac{(\tilde{c} - \tilde{u})\Delta A + \Delta Q}{2\tilde{c}} \end{aligned} \right\} \qquad (5.25)$$

$$\left. \begin{aligned} \tilde{\lambda}_1 &= \tilde{u} - \tilde{c} \\[2mm] \tilde{\lambda}_2 &= \tilde{u} + \tilde{c} \end{aligned} \right\} \qquad (5.26)$$

$$\left. \begin{aligned} \tilde{\mathbf{K}}_1 &= [1, \tilde{u} - \tilde{c}]^T \\[2mm] \tilde{\mathbf{K}}_2 &= [1, \tilde{u} + \tilde{c}]^T \end{aligned} \right\} \qquad (5.27)$$

$$\left. \begin{aligned} \Delta A &= A^+ - A^- \\[2mm] \Delta Q &= Q^+ - Q^- \end{aligned} \right\} \qquad (5.28)$$

$$\left. \begin{aligned} \tilde{u} &= \frac{Q^+\sqrt{A^-} + Q^-\sqrt{A^+}}{\sqrt{A^- A^+}\left(\sqrt{A^-} + \sqrt{A^+}\right)} \\[2mm] \tilde{c} &= \sqrt{\frac{g}{2}[(A/T)^- + (A/T)^+]} \end{aligned} \right\} \qquad (5.29)$$

Since the hydrostatic pressure and wall pressure terms are combined in Equation (5.12), the source terms must be discretized appropriately. Different source term treatment methods can be found in literature such as Garcia-Navarro and Vazquez-Cendon (2000), Perthame and Simeoni (2001), Ying et al. (2004), and Catella et al. (2008), among others. The treatment suggested by Lai and Khan (2012) is used to discretize the combined hydrostatic pressure force term. For linear elements, the source term associated with the water surface gradient is discretized as given by Equation (5.30).

$$
\left.
\begin{aligned}
-gN_1\hat{A}\frac{\partial \hat{Z}}{\partial x} &= -gN_1 \frac{(A_{x_s}^- + A_{x_e}^-)}{2}\frac{(Z_{x_e}^- - Z_{x_s}^-)}{(x_e - x_s)} \\[2mm]
-gN_2\hat{A}\frac{\partial \hat{Z}}{\partial x} &= -gN_2 \frac{(A_{x_s}^+ + A_{x_e}^+)}{2}\frac{(Z_{x_e}^+ - Z_{x_s}^+)}{(x_e - x_s)}
\end{aligned}
\right\}
\tag{5.30}
$$

## 5.3 Numerical tests

In this section, numerical solutions for the shallow water flow equations with the discontinuous Galerkin method are presented in nonrectangular and nonprismatic channels. Numerical tests include an idealized dam break, partial dam break, dam break in converging/diverging channel, hydraulic jump in a divergent channel, and flow in natural rivers. The numerical results are compared with the analytical solutions, laboratory measurements, and field data.

### 5.3.1 Dam break in a horizontal channel with different cross-section shapes

The discontinuous Galerkin scheme for the shallow water flow equations is used to simulate idealized dam-break problems in frictionless, horizontal channels with different cross-section shapes. The shapes tested include triangular channel ($b = 2y$), parabolic channel $\left(b = \sqrt{y}\right)$, trapezoidal channel ($b = 1 + 4y$), and rectangular channel ($b = 1$). The channel is 100 m long with a dam located at 50 m. Numerical tests for different cross-section shapes have the same initial conditions, that is, still water with water depth of 1 m upstream of the dam and water depth of 0.1 m downstream of the dam. The computational domain is discretized with 200 elements, and a time step of 0.02 s is used for all four tests. The exact and numerical solutions are compared in Figures 5.2 to 5.5. The exact solution can be found in Henderson (1966). The comparisons of the numerical and exact solutions show that the discontinuous Galerkin scheme is capable of simulating a shock wave in channels of different cross-section shapes.

### 5.3.2 Hydraulic jump in a divergent channel

In this test, a hydraulic jump in a divergent channel is simulated. The numerical results are compared with the measured water surface profile

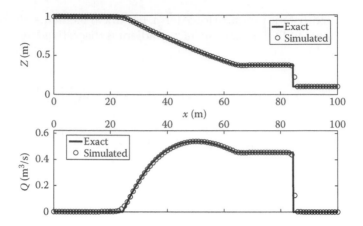

**Figure 5.2** Comparison of exact and numerical solutions for the idealized dam break in the parabolic channel.

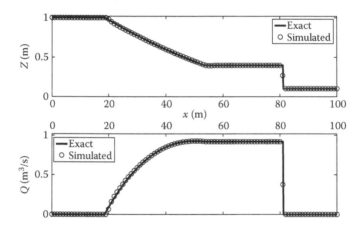

**Figure 5.3** Comparison of exact and numerical solutions for the idealized dam break in the rectangular channel.

(Khalifa, 1980). The divergent channel is a 2.5 m long horizontal channel with a rectangular cross-section. The cross-section width (in meters) along the channel is given by Equation (5.31). The width of the rectangular channel changes from 0.155 m to 0.46 m linearly over a length of 1.3 m.

$$b(x) = \begin{cases} 0.155 & \text{if} & 0 \le x \le 0.65 \\ 0.155 + 0.236(x - 0.65) & \text{if} & 0.65 < x < 1.94 \\ 0.46 & \text{if} & 1.94 \le x \le 2.5 \end{cases} \qquad (5.31)$$

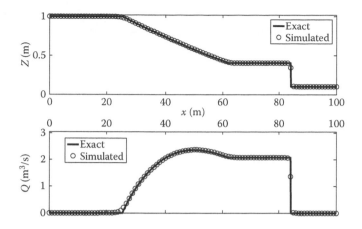

**Figure 5.4** Comparison of exact and numerical solutions for the idealized dam break in the trapezoidal channel.

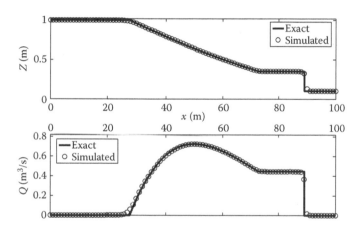

**Figure 5.5** Comparison of exact and numerical solutions for the idealized dam break in the triangular channel.

The initial water depth is 0.088 m throughout the channel. The upstream inflow rate is set at 0.0263 m³/s and the upstream water depth is maintained at 0.088 m. These conditions establish a supercritical flow condition at the upstream end. The water depth at the downstream end is increased from 0.088 m to 0.195 m in 50 seconds and is kept constant thereafter at 0.195 m. The Manning roughness coefficient is unknown, so simulations are performed with different values of roughness coefficient. In Figure 5.6, the steady solutions are shown for the water surface and discharge with a Manning roughness coefficient of 0.009. The location of the jump is predicted accurately and the mass flow rate is conserved.

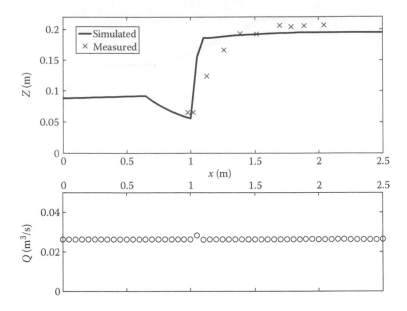

**Figure 5.6** Hydraulic jump simulation in the divergent channel.

### 5.3.3  Waterways Experiment Station dam-break tests

A series of hydraulic experiments were conducted at the Waterways Experiment Station (WES) by the U.S. Army Corps of Engineers (1960) to investigate the flow resulting from fully breached and partially breached dam events in a rectangular channel. The measured data from these experiments are useful to compare the accuracy of the numerical model for real-world dam-break problems. The test flume (shown in Figure 5.7) was 121.92 m (400 ft) long, 1.2192 m (4 ft) wide, with a bed slope of 0.005. The dam, located in the middle of the flume, was 0.3048 m (1 ft) high and water was ponded to the top of the dam in the upstream section. The flume bed downstream of the dam was dry. The same initial conditions are used in the numerical simulation. The Manning roughness coefficient

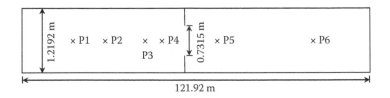

**Figure 5.7** Channel layout for partially and fully breached dam-break tests.

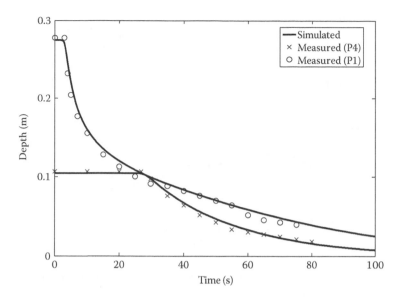

*Figure 5.8* Simulated and measured hydrographs at P1 and P4 for the fully breached dam-break test.

of 0.009 s/m$^{1/3}$ is used in the simulation. The cases of sudden removal of the dam and a partial dam break with a 0.7315 m (2.4 ft) wide breach are considered in this test.

The simulated hydrographs at the locations shown in Figure 5.7 are compared with the available measured data. For the fully breached case, the simulated and measured hydrographs at P1 (39.624 m upstream of the dam) and P4 (5.7912 m upstream of the dam) are shown in Figure 5.8, while the hydrographs at P5 (7.62 m downstream of dam) and P6 (45.72 m downstream of dam) are shown in Figure 5.9. For the partial dam-break case with a 0.7315 m wide breach, the simulated and measured hydrographs at P2 (30.68 m upstream of dam) and P3 (6.09 m upstream of dam) are shown in Figure 5.10, and the hydrographs at P5 and P6 are shown in Figure 5.11. For both tests, 121 elements are used, which gives an element size of 1 m. The time step size of 0.005 s is used in the simulations. The simulated results are in good agreement with the measured data.

### 5.3.4  Dam-break tests in a converging/diverging channel

A series of experiments were conducted in a converging/diverging rectangular channel by Bellos et al. (1992) to investigate the movement of a two-dimensional flood wave after an instantaneous dam break. The channel

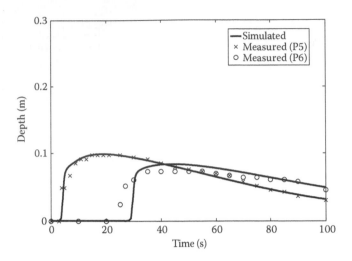

*Figure 5.9* Simulated and measured hydrographs at P5 and P6 for the fully breached dam-break test.

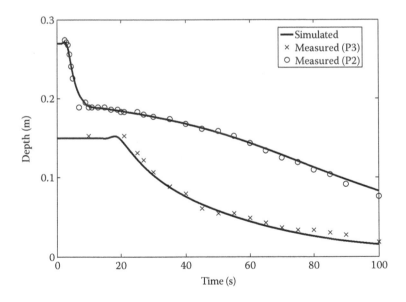

*Figure 5.10* Simulated and measured hydrographs at P2 and P3 for the partially breached dam-break test.

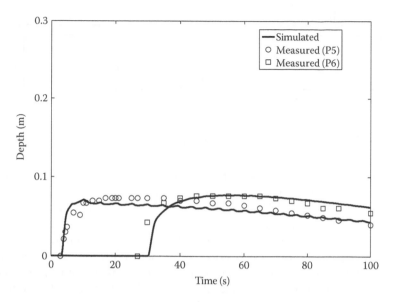

**Figure 5.11** Simulated and measured hydrographs at P5 and P6 for the partially breached dam-break test.

geometry used in the experiments is given in Table 5.1 and is shown in Figure 5.12. A gate was installed at the minimum width location in the channel ($x = 8.5$ m) to model the dam. The authors carried out various experiments with a different bed slope under both wet and dry bed conditions. In this test, dam-break flows for both wet bed and dry bed conditions downstream of the dam over a horizontal bed are simulated. The water depth is set at 0.3 m upstream of the gate for both tests, and the water depth is 0.101 m downstream of the gate for the wet bed dam break to mimic the experimental setup. The dam is removed instantaneously. The Manning roughness coefficient of 0.012 s/m$^{1/3}$ is used in the test. An element size of 0.1 m is used for both cases, and the time steps of 0.003 s and 0.0002 s are used for the wet bed and dry bed cases, respectively. Numerical solutions

**Table 5.1** Channel Width Variation for a Converging/Diverging Channel

| $x$ (m) | 0.0 | 5.0 | 5.5 | 6.0 | 6.5 | 7.0 | 7.5 | 8.0 | 8.5 |
|---|---|---|---|---|---|---|---|---|---|
| $b$ (m) | 1.40 | 1.40 | 1.22 | 1.05 | 0.90 | 0.77 | 0.67 | 0.62 | 0.60 |
| $x$ (m) | 9.0 | 9.5 | 10.0 | 10.5 | 11.0 | 11.5 | 12.0 | 12.5 | 13.0 |
| $b$ (m) | 0.61 | 0.62 | 0.64 | 0.68 | 0.75 | 0.82 | 0.91 | 0.99 | 1.08 |
| $x$ (m) | 13.5 | 14.0 | 14.5 | 15.0 | 15.5 | 16.0 | 16.5 | 18.0 | 21.2 |
| $b$ (m) | 1.15 | 1.24 | 1.28 | 1.33 | 1.37 | 1.39 | 1.40 | 1.40 | 1.40 |

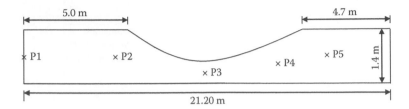

**Figure 5.12** Converging/diverging channel geometry.

are compared with the measured data at the gauge points where measurements are available. Gauge points are denoted as P1 ($x = 0$ m), P2 ($x = 4.5$ m), P3 ($x = 8.5$ m), P4 ($x = 13.5$ m), and P5 ($x = 18.5$ m) in Figure 5.12.

For the dam-break flow over an initially dry bed, the simulated and measured hydrographs at gauge points P1 and P5 are shown in Figure 5.13. The numerical solutions are in good agreement with the measured data. The simulated hydrograph at P5 is higher than the measured values as the dam break is assumed to be instantaneous, while in reality, the removal of the gate takes some time. For the dam-break flow over a wet bed test, a weir was installed at the downstream end of the channel. The configuration of the weir at the downstream end is not available. Since the downstream boundary condition is crucial to the numerical simulation, this may greatly impact the accuracy of the numerical solution. For the wet

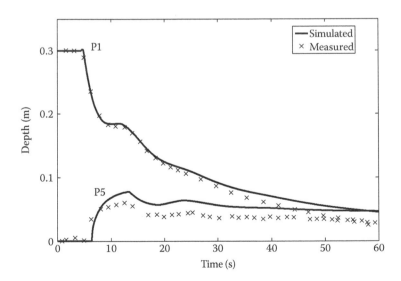

**Figure 5.13** Simulated and measured hydrographs for the dam-break test over a dry bed in the converging/diverging channel.

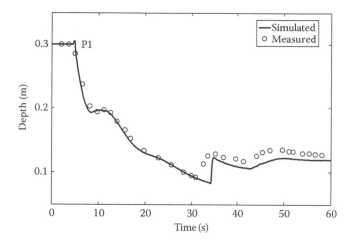

*Figure 5.14* Simulated and measured hydrographs at P1 for the dam-break test over a wet bed in the converging/diverging channel.

bed test, a rectangular sharp-crested weir is assumed at the downstream end of the channel in the simulation. The discharge relationship (Bazin's formula) is given by Equation (5.32), where the discharge coefficient ($C_d$) is taken as 0.62, and the length of the weir is equal to the channel width ($b = 1.4$ m).

$$Q = \frac{2}{3}C_d b(2g)^{2/3}(Z - 0.101)^{1.5} \tag{5.32}$$

The simulated and measured data for 5 gauge points are shown in Figures 5.14 to 5.18. The simulated hydrograph are in good agreement with the experimental measurement. The differences mostly come from the effect of the unknown weir structure.

### 5.3.5 *Teton Dam break flood simulation*

The Teton Dam site was located in the Teton River canyon (Idaho). The failure of the Teton Dam occurred on June 5, 1976, producing a significant flood downstream. The Teton Dam was a 92.96 m (305 ft) high earth filled dam with a crest length of 914.4 m (3000 ft). The inundated area and measured cross-sections after the dam break are shown in Figure 5.19. The measurement of the river cross-sections, axis along the river, the Manning roughness coefficient, reservoir storage depletion, and discharge at the dam site were documented by the U.S. Geological Survey (Ray and Kjelstrom, 1976). The cross-sections used in the computation are interpolated from

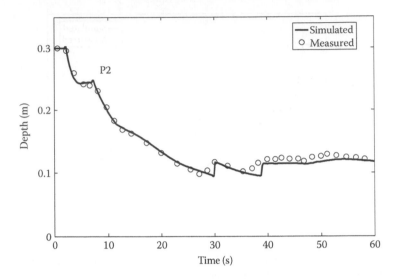

***Figure 5.15*** Simulated and measured hydrographs at P2 for the dam-break test over a wet bed in the converging/diverging channel.

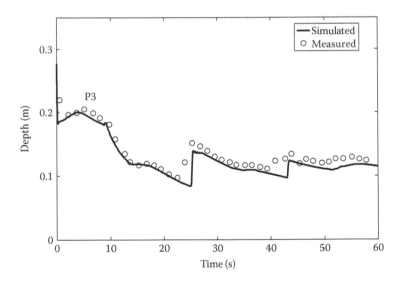

***Figure 5.16*** Simulated and measured hydrographs at P3 for the dam-break test over a wet bed in the converging/diverging channel.

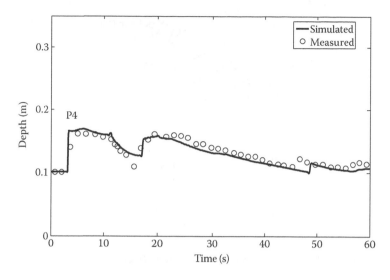

*Figure 5.17* Simulated and measured hydrographs at P4 for the dam-break test over a wet bed in the converging/diverging channel.

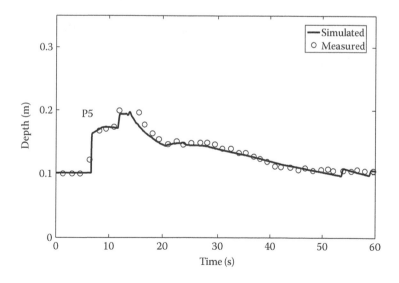

*Figure 5.18* Simulated and measured hydrographs at P5 for the dam-break test over a wet bed in the converging/diverging channel.

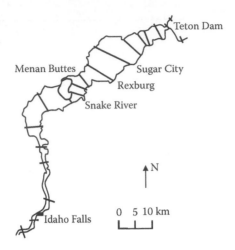

**Figure 5.19** Flood area downstream of the Teton River.

the available data. During the computation, the depth–area ($h$–$A$) relation, depth–wetted perimeter ($h$–$P_w$) relation, and depth–top width ($h$–$T$) relation are needed. These relationships are established for each cross-section. The discharge at the dam site after the dam failure is shown in Figure 5.20. The discharge and water surface level at the dam site are used as inflow boundary conditions if the inlet Froude number is larger than unity (supercritical inflow condition), otherwise only the discharge is used as an inflow boundary condition. Since the initial flow conditions before the flood are not available, a downstream dry bed condition is used in the simulation.

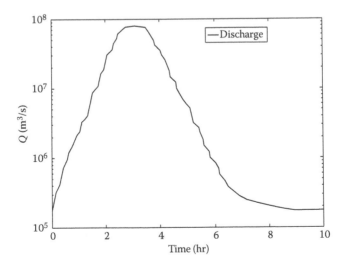

**Figure 5.20** Discharge at the Teton Dam site after the dam break.

In addition, the lateral inflow from Henry's Fork and the Snake River are ignored as these are insignificant compared to the flow from the dam. The simulation lasts for a period of 10 hours after the dam break.

The computed water surface levels and Froude numbers along the river at 8 hours after the dam break are shown in Figures 5.21 and 5.22,

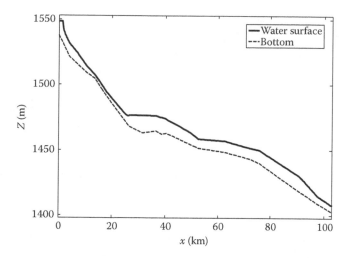

*Figure 5.21* Simulated water surface level at 8 hours after the Teton Dam break.

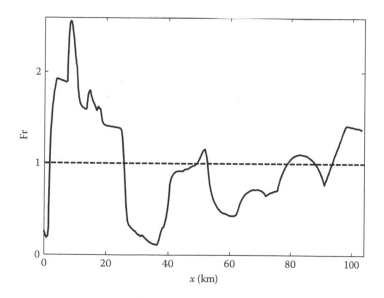

*Figure 5.22* Simulated Froude number along the river at 8 hours after the Teton Dam break.

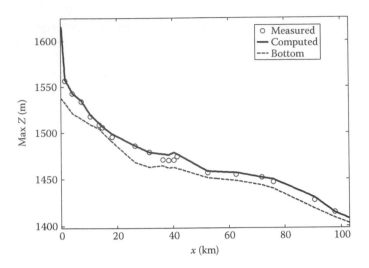

**Figure 5.23** Computed and measured maximum water elevation after the Teton Dam break.

respectively. The results show reasonable variation along the river. The comparison of the simulated maximum water surface elevations during the 10 hour flood event with the measured data is shown in Figure 5.23. The difference between the simulated results and measurement in the middle of the river (35–45 km) is mainly due to the omission of the lateral inflow from the Snake River.

### 5.3.6   The Toce River test

The Toce River physical model test used in the Concerted Action on Dambreak Modeling (CADAM) project (Frazão and Testa, 1999) is considered here. The physical model was a 1:100 scale model of a reach of the Toce River valley (Northern Alps, Italy) developed at the ENEL-HYDRO (Ente Nazionale per l'Energia Elettrica) laboratory in Milan, Italy. Details of the modeling parameters, such as the topographic data, the Manning roughness coefficient, and inflow discharge, were specified by Electricité de France (EDF). Measurements made during the physical model tests were also provided so that numerical modelers could make an objective evaluation of the simulation results.

The topography of the Toce River physical model covered approximately an area 50 m × 12 m and is shown in Figure 5.24. A rectangular tank was located at the upstream end of the physical model. Water was released from the tank to simulate the dam-break flood. A number

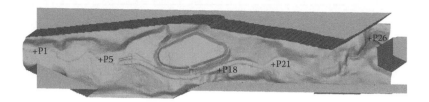

*Figure 5.24* Plan view of the Toce River.

of gauges were installed in the physical model to measure the water surface level after the release of water at the upstream end of the channel. Selected gauges along the main river axis (P1, P5, P18, P21, and P26) are used to compare the simulation results with the measured data. The computational mesh with 62 cross-sections used in this simulation is shown in Figure 5.25.

A dam-break flood occurred after the release of water from the rectangular tank at the upstream end of the channel. The measured discharge from the rectangular tank is used as the inflow boundary condition with critical flow. The discharge hydrograph at the rectangular tank is shown in Figure 5.26. The river basin is initially dry. A critical flow boundary condition is also applied at the outlet of the river reach as was the case in the physical model test.

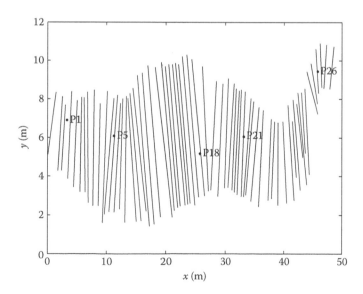

*Figure 5.25* Computational mesh and cross-sections of the Toce River.

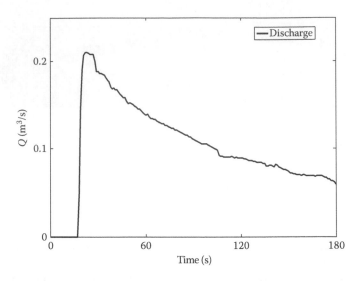

***Figure 5.26*** Inflow hydrograph at the upstream boundary of the Toce River.

A Manning roughness coefficient of 0.0162 s/m$^{1/3}$ is used based on the value proposed in the physical model study. The simulation period is 180 seconds after the release of water. The computed water surface levels and Froude numbers along the river axis at 100 seconds are shown in Figures 5.27 and 5.28, respectively. Several hydraulic jumps can be identified from the computed Froude number variation shown in Figure 5.28.

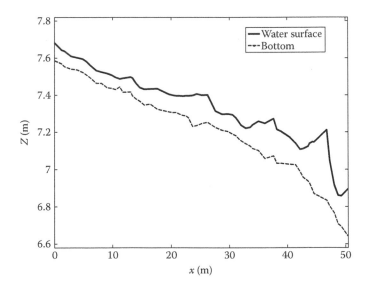

***Figure 5.27*** Simulated water surface profile along the Toce River at 100 s.

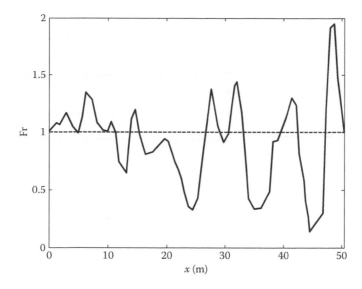

*Figure 5.28* Computed Froude number along the Toce River at 100 s.

The computed maximum water surface level during the dam-break flood is compared with the measured data in Figure 5.29. The computed and measured stage-time hydrographs at gauge points are compared in Figure 5.30. The simulated results of maximum water surface level and hydrographs are in good agreement with the measured data.

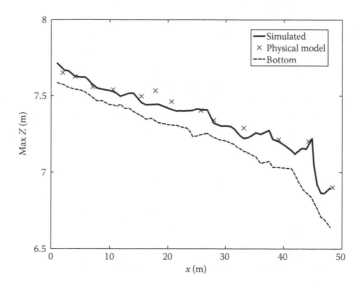

*Figure 5.29* Simulated maximum water surface profile along the Toce River.

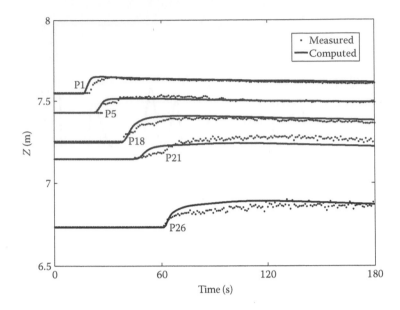

**Figure 5.30** Computed and measured hydrographs in the Toce River.

### 5.3.7  The East Fork River test

The natural river flow in the East Fork River is simulated with the one-dimensional model. The East Fork River is in the Wind River Range of Wyoming, west of the continental divide and east and southeast of Mount Bonneville. The river reach is approximately 3.3 km in length and is shown in Figure 5.31. The meandering study reach terminates at a bed-load trap constructed across the river. The numbers shown at each cross-section is the centerline distance (in meters) upstream from the bedload trap (section 0000). Field measurements of topography at 39 cross-sections, inflow rate, outflow rate, and hydrograph at the bedload trap are available (Emmett et al., 1980). This data is used in the simulation. The influence of the sediment transport and morphological changes are not considered in the simulation.

The river flow is simulated for a period of 12 days from June 1 to June 12, 1979. The average water surface level on June 1 is used as an initial condition with the initial flow rate of 6.0 m$^3$/s throughout the river reach. The hourly discharge measured at section 3295 is used as the inflow boundary condition with a subcritical flow. The hourly gage height at section 0000 is used as the outflow boundary condition. Since

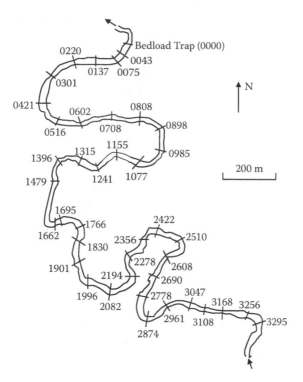

*Figure 5.31* Map of the study reach in the East Fork River (Emmett et al., 1980).

the roughness coefficient is unavailable, numerical tests with different Manning roughness coefficients are conducted to find the best fit value. The results presented are with the Manning roughness coefficient of 0.028.

The simulated and measured discharges at the outflow section 0000 are shown in Figure 5.32. The simulated and measured water surface elevations at section 2505 and section 3295 are shown in Figure 5.33. In Figure 5.34, the simulated and measured water surface levels at noon on June 12, 1979 are presented. In general, the simulated results are in good agreement with the field measurements. In Figure 5.33, the difference between simulated and measured data at the beginning may be due to the inaccuracy of the initial condition. The difference during the last 4 days may be due to the high sediment transport rate occurring during peak flow.

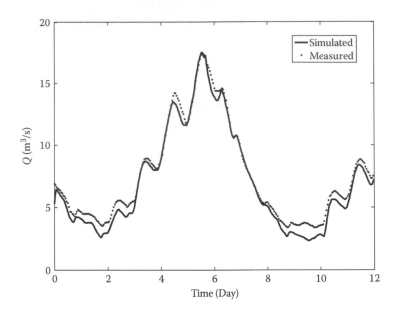

***Figure 5.32*** Computed and measured discharges at section 0000 of the East Fork River.

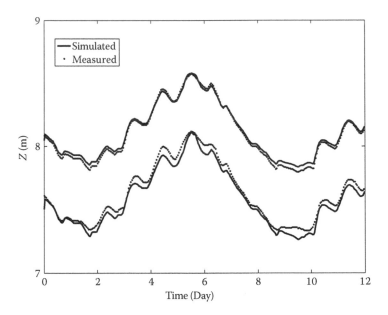

***Figure 5.33*** Computed and measured water surface profiles at section 3295 (upper lines) and section 2505 (lower lines) of the East Fork River.

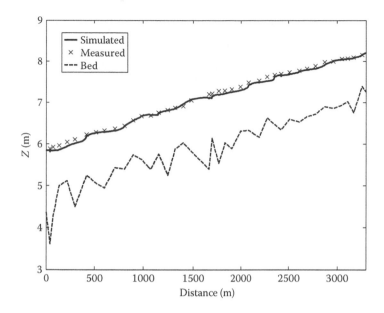

**Figure 5.34** Computed and measured water surface profiles around noon on June 12, 1979, in the East Fork River.

# References

Bellos, C., Soulis, V., and Sakkas, J. (1992). Experimental investigation of two-dimensional dam-break induced flows. *Journal of Hydraulic Research*, 30(1), 47–63.

Catella, M., Paris, E., and Solari, L. (2008). Conservative scheme for numerical modeling of flow in natural geometry. *Journal of Hydraulic Engineering*, 134(6), 736–748.

Cunge, J. A., Holly, F. M., and Verwey, A. (1980). *Practical Aspects of Computational River Hydraulic*. Pitman, London.

Emmett, W. M., Myrick, R. M., and Meade, R. H. (1980). Field data describing the movement and storage of sediment in the East Fork River, Wyoming, Part I. River hydraulics and sediment transport, 1979. USGS Open-File Report 80-1189, Denver.

Fraccarollo, L., and Toro, E. F. (1995). Experimental and numerical assessment of the shallow water model for two-dimensional dam-break type problems. *Journal of Hydraulic Research*, 33(6), 843–864.

Frazão, S. S., and Testa, G. (1999). The Toce River test case: Numerical results analysis. Proceedings of the 3rd CADAM Workshop, Milan, Italy.

Garcia-Navarro, P., and Vazquez-Cendon, M. E. (2000). On numerical treatment of the source terms in the shallow water equations. *Computers & Fluids*, 29(8), 951–979.

Henderson, F. M. (1966). *Open Channel Flow*. McGraw-Hill, New York.

Khalifa, A. M. (1980). Theoretical and experimental study of the radial hydraulic jump. Ph.D. thesis, University of Windsor, Windsor, Ontario, Canada.

Lai, W., and Khan, A. A. (2012). Discontinuous Galerkin method for 1D shallow water flow in non-rectangular and non-prismatic channels. *Journal of Hydraulic Engineering*, 138(3), 285–296.

Perthame, B., and Simeoni, C. (2001). A kinetic scheme for the Saint-Venant system with a source term. *Calcolo*, 38(4), 201–231.

Ray, H. A., and Kjelstrom, L. C. (1976). The flood in southeastern Idaho from the Teton dam failure of June 5, 1976. U.S. Geological Survey, Open-File Report 77-765, Boise, Idaho.

U.S. Army Corps of Engineers. (1960). Floods resulting from suddenly breached dams: Conditions of minimum resistance. Miscellaneous paper No. 2-374, Report 1, Waterways Experiment Station, Vicksburg, Mississippi.

Ying, X., Khan, A. A., and Wang, S. S. Y. (2004). Upwind conservative scheme for the Saint Venant equations. *Journal of Hydraulic Engineering*, 130(10), 977–987.

*chapter six*

# Two-dimensional conservation laws

Formulations of the discontinuous Galerkin (DG) method for a two-dimensional (2D) scalar equation and equations representing a system of conservation laws in two dimensions are presented in this chapter. The treatment of flux terms is discussed. The implementation of different slope limiters is shown. Details of the numerical solution with the discontinuous Galerkin method for pure convection problems and shallow water flow equations in 2D are discussed.

## 6.1 Pure convection in 2D

In this section, the governing equation and properties of pure convection in 2D are discussed. The details of the numerical solution with the discontinuous Galerkin method are provided. Numerical tests are conducted for linear, pure convection problems. The application of slope limiters and approximation of the flux term are also shown.

### 6.1.1 Governing equation of convection in 2D

The governing equation of pure convection in 2D can be written in scalar conservation form as given by Equation (6.1). The flux term is given by Equation (6.2). In these equations, $u$ and $v$ are the velocities in $x$ and $y$ directions, respectively, and $C$ is the concentration of the pollutant in the flow. The eigenvalue of the 2D convection equation is given by Equation (6.3), where $\mathbf{n} = (n_x, n_y)$ is a unit normal vector and $u_n$ is defined as the normal velocity in the direction of $\mathbf{n}$. It is obvious that the governing equation is hyperbolic.

$$\frac{\partial C}{\partial t} + \nabla \cdot \mathbf{F} = \frac{\partial C}{\partial t} + \frac{\partial uC}{\partial x} + \frac{\partial vC}{\partial y} = 0 \tag{6.1}$$

$$\mathbf{F} = (E, G) = (uC, vC) \tag{6.2}$$

$$\lambda = \frac{\partial \mathbf{F} \cdot \mathbf{n}}{\partial C} = un_x + vn_y \triangleq u_n \tag{6.3}$$

## 6.2   Discontinuous Galerkin formulation for 2D convection

For the discontinuous elements, unknown $C$ inside an element is approximated using polynomial shape functions as given by Equation (6.4). In Equation (6.5), the governing equation is multiplied with the weight function, $N_i(x, y)$, and integrated over an element to obtain the weighted residual formulation. The flux term is integrated using Gauss theorem, resulting in Equation (6.6).

$$
\left.
\begin{aligned}
C \approx \hat{C} = \sum N_j(x,y)C_j \\
\mathbf{F} \approx \hat{\mathbf{F}} = \mathbf{F}(\hat{C})
\end{aligned}
\right\}
\tag{6.4}
$$

$$
\int_{\Omega_e} N_i \frac{\partial C}{\partial t} d\Omega + \int_{\Omega_e} N_i \nabla \cdot \mathbf{F} d\Omega = 0
\tag{6.5}
$$

$$
\int_{\Omega_e} N_i N_j \, d\Omega \frac{\partial C_j}{\partial t} + \int_{\Gamma_e} N_i \tilde{\mathbf{F}} d\Gamma - \int_{\Omega_e} (\nabla N_i) \cdot \hat{\mathbf{F}} d\Omega = 0
\tag{6.6}
$$

The numerical flux can be calculated with the upwind scheme given by Equation (6.7). The wave velocity at the boundary is approximated using Equation (6.8). In these equations, $\mathbf{n}$ is the outward unit normal vector at an element boundary, $u_{nL}$ and $u_{nR}$ are the normal velocities at the left and right sides of the element boundary, and $C_L$ and $C_R$ are the pollutant quantity at the left and right sides of the element boundary (the velocities and the pollutant quantities are based on the averaged value of the boundary nodes). For a triangle $\Delta_{123}$ (shown in Figure 6.1), the nodes are numbered in the counterclockwise direction. The left side of the boundary is always referred to as the triangle $\Delta_{123}$, that is, the element under consideration, and the right side of the boundary is the neighboring element.

$$
\tilde{\mathbf{F}} =
\begin{cases}
\mathbf{F}_L \cdot \mathbf{n} = u_{nL} C_L & \text{if} \quad \lambda \geq 0 \\
\mathbf{F}_R \cdot \mathbf{n} = u_{nR} C_R & \text{if} \quad \lambda < 0
\end{cases}
\tag{6.7}
$$

$$
\lambda = \frac{1}{2}(u_{nL} + u_{nR})
\tag{6.8}
$$

Numerical integration of Equation (6.6) can be carried out with isoparametric mapping and transformation from the global to local coordinates as shown in Equation (6.9). The explanation of the individual terms is given in Equations (6.10) to (6.12). Further explanation of Equation (6.12) is given in Chapter 2 by Equations (2.49) to (2.51). The

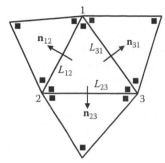

*Figure 6.1* Discontinuous linear triangular elements.

determinant of the Jacobian is equal to twice the area of the triangle. The final equation can be written in a compact form as shown in Equation (6.13).

$$\int_0^1 \int_0^1 N_i N_j \det(\mathbf{J}^e) d\xi\, d\eta \frac{\partial C_j}{\partial t} + \sum_{j=1,\, j\neq i}^{3} \tilde{\mathbf{F}}_{ij} \frac{L_{ij}}{2} -$$
$$\int_0^1 \int_0^1 \nabla' \mathbf{N}_i \bullet \mathbf{F}(\hat{\mathbf{U}}) \det(\mathbf{J}^e) d\xi\, d\eta = 0 \tag{6.9}$$

$$\int_0^1 \int_0^1 N_i N_j \det(\mathbf{J}^e) d\xi\, d\eta \frac{\partial C_j}{\partial t} = \frac{Area}{12} \begin{bmatrix} 2 & 1 & 1 \\ 1 & 2 & 1 \\ 1 & 1 & 2 \end{bmatrix} \frac{\partial}{\partial t} \begin{bmatrix} C_1 \\ C_2 \\ C_3 \end{bmatrix} \tag{6.10}$$

$$\sum_{j=1,\, j\neq i}^{3} \tilde{\mathbf{F}}_{ij} \frac{L_{ij}}{2} = \frac{1}{2} \begin{bmatrix} \tilde{\mathbf{F}}_{12}L_{12} + \tilde{\mathbf{F}}_{13}L_{13} \\ \tilde{\mathbf{F}}_{23}L_{23} + \tilde{\mathbf{F}}_{12}L_{12} \\ \tilde{\mathbf{F}}_{31}L_{31} + \tilde{\mathbf{F}}_{12}L_{12} \end{bmatrix} \tag{6.11}$$

$$\int_0^1 \int_0^1 \nabla' \mathbf{N}_i \bullet \mathbf{F}(\hat{\mathbf{U}}) \det(\mathbf{J}^e) d\xi\, d\eta = \int_0^1 \int_0^1 \left[ y_{31} \begin{bmatrix} -1 \\ 1 \\ 0 \end{bmatrix} - y_{21} \begin{bmatrix} -1 \\ 0 \\ 1 \end{bmatrix} \right] \hat{E}\, d\xi\, d\eta +$$

$$\int_0^1 \int_0^1 \left[ -x_{31} \begin{bmatrix} -1 \\ 1 \\ 0 \end{bmatrix} + x_{21} \begin{bmatrix} -1 \\ 0 \\ 1 \end{bmatrix} \right] \hat{G}\, d\xi\, d\eta \tag{6.12}$$

$$\mathbf{M} \frac{\partial \mathbf{C}}{\partial t} = \mathbf{R} \quad \text{or} \quad \frac{\partial \mathbf{C}}{\partial t} = \mathbf{M}^{-1}\mathbf{R} = \mathbf{L} \tag{6.13}$$

## 6.3   Slope limiters

The slope limiters are required to reduce spurious oscillation around shock waves, and maintain high-order accuracy in the smooth regions. There are many slope limiters available, such as minmod-type limiters (Burbeau et al., 2001), moment-based limiters (Biswas et al., 1994), monotonicity-preserving limiters (Suresh and Huynh, 1997), and weighted essentially nonoscillatory limiters (Qiu and Shu, 2005). Many slope limiters use Legendre polynomials, spectral orthogonal, or hierarchical basis functions for high-order constructions. For illustration purposes, the slope limiters for the second-order linear triangular elements are presented next. A typical linear triangular element is shown in Figure 6.2.

Two slope limiters are presented. The first slope limiter (SL1) is adopted from Lai and Khan (2012). The slope limiter is applied to the conserved variables. The limiting procedure includes the following four steps. In the first step, the average solution of the conserved variable is computed for the main element and the surrounding three elements ($e = 0, 1, 2, 3$) as given by Equation (6.14). In the second step, the unlimited gradients for all elements are calculated using Green's theorem. An example of the evaluation of the unlimited gradient for the element 0 is given by Equation (6.15).

$$\bar{U}_e = \frac{1}{3}\sum_{j=1}^{3}U_{e,j}, \quad U \in \mathbf{U} \tag{6.14}$$

$$
\left.
\begin{aligned}
\frac{\partial U}{\partial x} &= \frac{1}{\Omega_0}\oint_{\Gamma_0}U\,dy = \frac{1}{\Omega_0}\sum_{e=1}^{3}U_{0e}\Delta y_e \\[2pt]
&= \frac{[U_{01}(y_{n2}-y_{n1})+U_{02}(y_{n3}-y_{n2})+U_{03}(y_{n1}-y_{n3})]}{\Omega_0} \\[8pt]
\frac{\partial U}{\partial y} &= \frac{-1}{\Omega_0}\oint_{\Gamma_0}U\,dx = \frac{-1}{\Omega_0}\sum_{e=1}^{3}U_{0e}\Delta x_e \\[2pt]
&= \frac{[U_{01}(x_{n2}-x_{n1})+U_{02}(x_{n3}-x_{n2})+U_{03}(x_{n1}-x_{n3})]}{-\Omega_0}
\end{aligned}
\right\} \tag{6.15}
$$

In Equation (6.15), $U_{01}$ is the value at the boundary between elements 0 and 1. First, the average values in the neighboring elements are calculated. Then, the inverse distance weighting is used to find $U_{01}$ at the boundary. Similar procedure is adopted to calculate $U_{02}$ and $U_{03}$. In the third step, the limited gradient is computed by taking the weighted average of the unlimited gradient of the surrounding elements as shown in Equation (6.16). The weight factors in Equation (6.16) are given by Equation (6.17), where $\delta$ is a

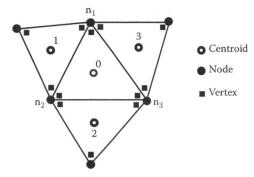

**Figure 6.2** Discretization using triangular elements in the DG method.

small number introduced to prevent indeterminacy of the weights, and $g_e$ ($e = 1, 2, 3$) for the surrounding elements is defined by Equation (6.18). Finally, the nodal values of the limited conservative variables ($U_{e,j}^l$) can be computed using Equation (6.19).

$$(\nabla U)_0^l = w_1(\nabla U)_1 + w_2(\nabla U)_2 + w_3(\nabla U)_3 \tag{6.16}$$

$$\left.\begin{aligned} w_1 &= \frac{g_2 g_3 + \delta}{g_1^2 + g_2^2 + g_3^2 + 3\delta} \\[2ex] w_2 &= \frac{g_1 g_3 + \delta}{g_1^2 + g_2^2 + g_3^2 + 3\delta} \\[2ex] w_3 &= \frac{g_1 g_2 + \delta}{g_1^2 + g_2^2 + g_3^2 + 3\delta} \end{aligned}\right\} \tag{6.17}$$

$$\left.\begin{aligned} g_1 &= \left\|(\nabla U)_1\right\|^2 \\[1ex] g_2 &= \left\|(\nabla U)_2\right\|^2 \\[1ex] g_3 &= \left\|(\nabla U)_3\right\|^2 \end{aligned}\right\} \tag{6.18}$$

$$\left.\begin{aligned} \left(\frac{\partial U}{\partial x}\right)_0^l &= \sum_{j=1}^3 \frac{\partial N_j}{\partial x} U_{0,j}^l \\[2ex] \left(\frac{\partial U}{\partial y}\right)_0^l &= \sum_{j=1}^3 \frac{\partial N_j}{\partial y} U_{0,j}^l \\[2ex] \bar{U}_0 &= \frac{1}{3}\sum_{j=1}^3 U_{0,j}^l \end{aligned}\right\} \tag{6.19}$$

The second slope limiter (SL2) is adopted from Anastasiou and Chan (1997). The variables are limited as given by Equation (6.20), where $\nabla U$ is the unlimited gradient given by Equation (6.15), **r** is a vector with origin at the centroid of the element extending to any point within the element, and $\phi$ is a chosen limiter given by Equation (6.21). To determine the nodal values of the conserved variables, Equation (6.20) is evaluated at the three nodes. The vector **r** for the node 1 will be $x_{m_1} - \bar{x}$, $y_{m_1} - \bar{y}$, where $\bar{x}$ and $\bar{y}$ are the values at the center of the element. The parameters in Equation (6.21) are explained in Equations (6.22) to (6.24), where $k = 1, 2, 3$ represents the sides of element 0. In the equations, $U_k$ is the unlimited cell edge value and $\bar{U}_e$ is the unlimited cell centroid value. $U_k$ is the average value at the boundary $k$, determined from the nodal values of element 0. The value of parameter $\beta$ ranges from 1 to 2, in particular $\beta = 1$ and $\beta = 2$ result in the minmod limiter and Superbee limiter, respectively. If $\phi$ is taken as zero, the scheme degenerates to the first-order finite volume method (FVM).

$$U_0^l(x,y) = \bar{U}_0 + \phi \nabla U \cdot \mathbf{r} \tag{6.20}$$

$$\phi = \min(\phi_k), \quad k = 1, 2, 3 \tag{6.21}$$

$$\phi_k = \max[\min(\beta r_k, 1), \min(r_k, \beta)] \tag{6.22}$$

$$r_k = \begin{cases} \left(U_0^{max} - \bar{U}_0\right)/\left(U_k - \bar{U}_0\right) & \text{if} \quad U_k - \bar{U}_0 > 0 \\ \left(U_0^{min} - \bar{U}_0\right)/\left(U_k - \bar{U}_0\right) & \text{if} \quad U_k - \bar{U}_0 < 0 \\ 1.0 & \text{if} \quad U_k - \bar{U}_0 = 0 \end{cases} \tag{6.23}$$

$$U_0^{min} = \min(\bar{U}_e), \quad U_0^{max} = \max(\bar{U}_e), \quad e = 0, 1, 2, 3 \tag{6.24}$$

## 6.4   Numerical tests

The numerical scheme is first applied to a pure convection problem involving a rotating cone-shaped scalar field. The initial condition is shown in Figure 6.3. The cone-shaped scalar has a maximum value of 1.0 unit and base radius of 0.1 m. The scalar field is convected by a stationary velocity field with angular velocity $\omega = 2.0$ rad/s. The simulated results after one rotation period are shown in Figures 6.4 to 6.6 with different limiters. The numerical results along the lines $y = 0.75$ and $x = 0.5$ are compared with exact solutions in Figure 6.7. The first-order FVM is diffusive, the DG

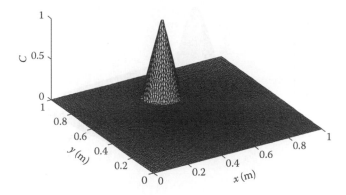

*Figure 6.3* Initial condition for the cone-shaped pure convection test.

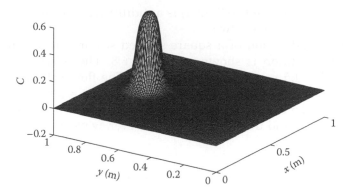

*Figure 6.4* Numerical results with SL1 for the cone-shaped initial condition.

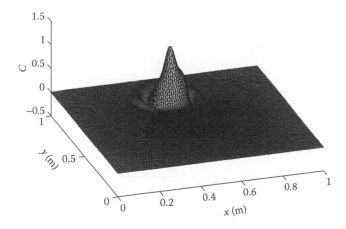

*Figure 6.5* Numerical results with SL2 for the cone-shaped initial condition.

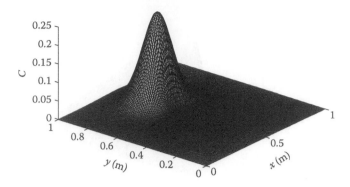

**Figure 6.6** Numerical results with the FVM for the cone-shaped initial condition.

method with SL2 (Superbee limiter) is antidiffusive, and the DG scheme with SL1 falls between the two.

Next, the advection of a square-shaped scalar field is simulated. The initial condition is shown in Figure 6.8. The initial size of the square field is 1.5 m × 1.5 m, with $C = 10$. The flow velocities are set to $u = 1$ m/s and $v = 1$ m/s. Simulated results at $t = 2$ s are shown in Figures 6.9 and 6.10. Slope limiter 1 is used in this test. Diffusion effects are observed around the square edge; however, wave movement is captured satisfactorily.

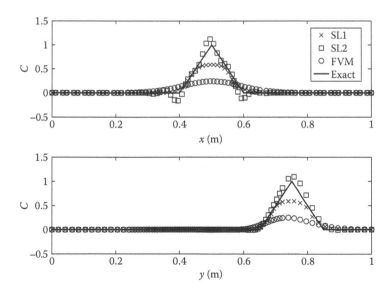

**Figure 6.7** Comparison of numerical (with different limiters) and exact solutions for the cone-shaped initial condition.

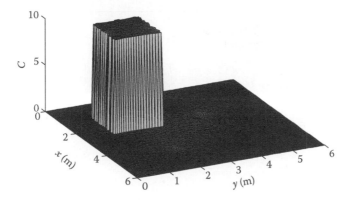

*Figure 6.8* Initial condition for the square-shaped pure convection test.

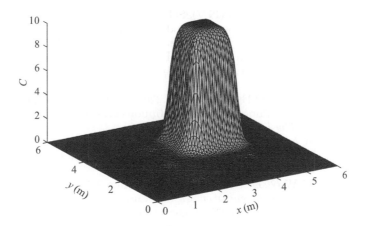

*Figure 6.9* Numerical results in 3D for the square-shaped initial condition at $t = 2$ s.

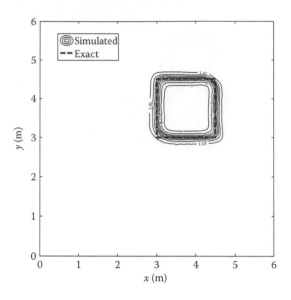

*Figure 6.10* Numerical results as contours for the square-shaped initial condition at $t = 2$ s.

## References

Anastasiou, K., and Chan, C. T. (1997). Solution of the 2D shallow water equations using the finite element method on unstructured triangular meshes. *International Journal for Numerical Methods in Fluids*, 24(11), 1225–1245.

Biswas, R., Devine, K. D., and Flaherty, J. E. (1994). Parallel adaptive finite element methods for conservation laws. *Applied Numerical Mathematics*, 14(1–3), 255–283.

Burbeau, A., Sagaut, P., and Hruneau, Ch.-H. (2001). A problem-independent limiter for high-order Runge–Kutta discontinuous Galerkin methods. *Journal of Computational Physics*, 169, 115–150.

Lai, W., and Khan, A. A. (2012). A discontinuous Galerkin method for two-dimensional shallow water flows. *International Journal for Numerical Methods in Fluids*, 70(8), 939–960.

Qiu, J., and Shu, C. W. (2005). Runge–Kutta discontinuous Galerkin method using WENO limiters. *SIAM Journal on Scientific Computing*, 26(3), 907–929.

Suresh, A., and Huynh, H. T. (1997). Accurate monotonicity-preserving schemes with Runge–Kutta time stepping. *Journal of Computational Physics*, 136(1), 83–99.

# chapter seven

# Two-dimensional shallow water flow in channels with horizontal beds

In this chapter, the discontinuous Galerkin (DG) method is applied to simulate the two-dimensional shallow water flows in channels with horizontal beds. The two-dimensional shallow water flow equations are introduced. Different treatments of numerical flux terms are discussed. The slope limiter presented by Lai and Khan (2012), as discussed in the previous chapter, is used for the numerical scheme. The numerical scheme is applied to a variety of test problems, and the numerical solutions are compared with the analytical solutions, laboratory measurements, and field data.

## 7.1 Two-dimensional shallow water flow equations for a horizontal bed channel

The two-dimensional shallow water flow equations are derived by integrating the 3D Navier–Stokes equations in the vertical direction under the assumptions of hydrostatic pressure distribution and uniform velocity in the vertical direction. The shallow water flow equations are applicable where the horizontal extent is much greater than the depth of flow and the vertical acceleration may be neglected. The depth-averaged shallow water flow equations are given by Equation (7.1). The friction slopes in the $x$ and $y$ directions are given by Equation (7.2).

$$
\left.
\begin{aligned}
\frac{\partial h}{\partial t} + \frac{\partial uh}{\partial x} + \frac{\partial vh}{\partial y} &= 0 \\[2mm]
\frac{\partial uh}{\partial t} + \frac{\partial (hu^2 + gh^2/2)}{\partial x} + \frac{\partial huv}{\partial y} &= -gh\frac{\partial z_b}{\partial x} - ghS_{fx} \\[2mm]
\frac{\partial vh}{\partial t} + \frac{\partial huv}{\partial x} + \frac{\partial (hv^2 + gh^2/2)}{\partial y} &= -gh\frac{\partial z_b}{\partial y} - ghS_{fy}
\end{aligned}
\right\}
\tag{7.1}
$$

$$S_{fx} = \frac{n^2 u \sqrt{u^2 + v^2}}{h^{4/3}}$$

$$\left.\begin{array}{c} \\ \\ \end{array}\right\}$$ (7.2)

$$S_{fy} = \frac{n^2 v \sqrt{u^2 + v^2}}{h^{4/3}}$$

The two-dimensional shallow water flow equations in the conservation form is given by Equation (7.3). The vectors of conserved variables, source vector, and flux vectors are given by Equation (7.4). To evaluate the property of the governing equations, the Jacobian matrix of the governing equations is analyzed. For an arbitrary unit vector $\mathbf{n} = (n_x, n_y)$, the Jacobian matrix of the governing equations is given by Equation (7.5). The Jacobian matrix has three distinct eigenvalues and eigenvectors as given by Equations (7.6) and (7.7), respectively. The eigenvalues show that the equations form a hyperbolic system.

$$\frac{\partial \mathbf{U}}{\partial t} + \nabla \cdot \mathbf{F}(\mathbf{U}) = \frac{\partial \mathbf{U}}{\partial t} + \frac{\partial \mathbf{E}(\mathbf{U})}{\partial x} + \frac{\partial \mathbf{G}(\mathbf{U})}{\partial y} = \mathbf{S}(\mathbf{U}) \qquad (7.3)$$

$$\mathbf{U} = \begin{pmatrix} h \\ hu \\ hv \end{pmatrix}, \mathbf{S} = \begin{pmatrix} 0 \\ -gh\dfrac{\partial z_b}{\partial x} - ghS_{fx} \\ -gh\dfrac{\partial z_b}{\partial y} - ghS_{fy} \end{pmatrix},$$

$$\mathbf{E}(\mathbf{U}) = \begin{pmatrix} hu \\ hu^2 + gh^2/2 \\ huv \end{pmatrix}, \mathbf{G}(\mathbf{U}) = \begin{pmatrix} hv \\ huv \\ hv^2 + gh^2/2 \end{pmatrix}$$

(7.4)

$$\mathbf{J}(\mathbf{U}) = \frac{\partial \mathbf{F}(\mathbf{U}) \cdot \mathbf{n}}{\partial \mathbf{U}} = \frac{\partial \mathbf{E}}{\partial \mathbf{U}} n_x + \frac{\partial \mathbf{G}}{\partial \mathbf{U}} n_y$$

$$= \begin{bmatrix} 0 & n_x & n_y \\ (gh - u^2)n_x - uvn_y & 2un_x + vn_y & un_y \\ -uvn_x + (gh - v^2)n_y & vn_x & un_x + 2vn_y \end{bmatrix} \qquad (7.5)$$

$$\left.\begin{aligned}
\lambda_1 &= un_x + vn_y - \sqrt{gh} \\
\lambda_2 &= un_x + vn_y \\
\lambda_3 &= un_x + vn_y + \sqrt{gh}
\end{aligned}\right\} \tag{7.6}$$

$$\mathbf{K}_1 = \begin{pmatrix} 1 \\ u - \sqrt{gh}n_x \\ v - \sqrt{gh}n_y \end{pmatrix}, \quad \mathbf{K}_2 = \begin{pmatrix} 0 \\ -n_y \\ n_x \end{pmatrix}, \quad \mathbf{K}_3 = \begin{pmatrix} 1 \\ u + \sqrt{gh}n_x \\ v + \sqrt{gh}n_y \end{pmatrix} \tag{7.7}$$

The shallow water flow equations can be written in an alternate form as given by Equation (7.8). The corresponding vector of conserved variables, source vector, and flux vectors are given by Equation (7.9). In these equations, $q_x$ and $q_x$ are discharges per unit width in the $x$ and $y$ directions and are equal to $uh$ and $vh$, respectively.

$$\left.\begin{aligned}
\frac{\partial h}{\partial t} + \frac{\partial q_x}{\partial x} + \frac{\partial q_y}{\partial y} &= 0 \\
\frac{\partial q_x}{\partial t} + \frac{\partial \left(q_x^2/h + gh^2/2\right)}{\partial x} + \frac{\partial q_x q_y/h}{\partial y} &= -gh\frac{\partial z_b}{\partial x} - ghS_{fx} \\
\frac{\partial q_y}{\partial t} + \frac{\partial q_x q_y/h}{\partial x} + \frac{\partial \left(q_y^2/h + gh^2/2\right)}{\partial y} &= -gh\frac{\partial z_b}{\partial y} - ghS_{fy}
\end{aligned}\right\} \tag{7.8}$$

$$\left.\begin{aligned}
\mathbf{U} = \begin{pmatrix} h \\ q_x \\ q_y \end{pmatrix}, \quad \mathbf{S} = \begin{pmatrix} 0 \\ -gh\dfrac{\partial z_b}{\partial x} - ghS_{fx} \\ -gh\dfrac{\partial z_b}{\partial y} - ghS_{fy} \end{pmatrix}, \\[2em]
\mathbf{E(U)} = \begin{pmatrix} q_x \\ q_x^2/h + gh^2/2 \\ q_x q_y/h \end{pmatrix}, \quad \mathbf{G(U)} = \begin{pmatrix} q_y \\ q_x q_y/h \\ q_y^2/h + gh^2/2 \end{pmatrix}
\end{aligned}\right\} \tag{7.9}$$

## 7.2   Numerical flux

The DG formulation for the two-dimensional flow system is similar to the two-dimensional scalar case discussed in the previous chapter. The DG formulation is applied to each equation of the system. For the 2D shallow water flow equations, the numerical flux can be calculated with the HLL (Harten-Lax-van Leer), HLLC (Harten-Lax-van Leer contact), or Roe flux functions discussed next. The HLL flux (Fraccarollo and Toro, 1995) and HLLC flux (Eskilsson and Sherwin, 2004) are given by Equations (7.10) and (7.11), respectively.

$$\mathbf{F}^{HLL} = \begin{cases} \mathbf{F}_L \cdot \mathbf{n} & \text{if} & S_L \geq 0 \\ \dfrac{(S_R \mathbf{F}_L - S_L \mathbf{F}_R) \cdot \mathbf{n} + S_L S_R (\mathbf{U}_R - \mathbf{U}_L)}{S_R - S_L} & \text{if} & S_L < 0 < S_R \\ \mathbf{F}_R \cdot \mathbf{n} & \text{if} & S_R \leq 0 \end{cases} \tag{7.10}$$

$$\mathbf{F}^{HLLC} = \begin{cases} \mathbf{F}(\mathbf{U}_L) \cdot \mathbf{n} & \text{if} & S_L \geq 0 \\ \mathbf{F}(\mathbf{U}_L) \cdot \mathbf{n} + S_L (\mathbf{U}_{*L} - \mathbf{U}_L) & \text{if} & S_L < 0 \leq S_* \\ \mathbf{F}(\mathbf{U}_R) \cdot \mathbf{n} + S_R (\mathbf{U}_{*R} - \mathbf{U}_R) & \text{if} & S_* < 0 < S_R \\ \mathbf{F}(\mathbf{U}_R) \cdot \mathbf{n} & \text{if} & S_R \leq 0 \end{cases} \tag{7.11}$$

For the 2D shallow water flow equations, with a rotation matrix $\mathbf{T}$ and its inverse matrix as shown in Equation (7.12), the flux can be written as given by Equation (7.13). For $\mathbf{Q} = \mathbf{TU}$, the numerical flux is given by Equation (7.14). The vectors $\mathbf{Q}$ and $\mathbf{E}(\mathbf{Q})$ are given by Equation (7.15). The HLL and HLLC fluxes are given by Equations (7.16) and (7.17), respectively. The wave speeds and the intermediate variables are given by Equations (7.18) to (7.23), where $u_t$ is the tangential velocity. The quantities $\mathbf{E}(\mathbf{Q}_L)$ and $\mathbf{E}(\mathbf{Q}_R)$ represent values at the left and right sides of the element boundary moving in the counterclockwise direction. The $\mathbf{E}(\mathbf{Q}_L)$ and $\mathbf{E}(\mathbf{Q}_R)$ values at the boundary are calculated based on the averaged values of $h$, $q_x$, and $q_y$ of the corresponding boundary nodes. Once $\tilde{E}$ is evaluated, the numerical flux can then be calculated using Equation (7.14). For evaluating the numerical flux using the HLL flux function, the form given by Equation (7.10) can be used directly, where the wave speeds are given by Equations (7.18) and (7.19).

$$\mathbf{T} = \begin{bmatrix} 1 & 0 & 0 \\ 0 & n_x & n_y \\ 0 & -n_y & n_x \end{bmatrix}, \quad \mathbf{T}^{-1} = \begin{bmatrix} 1 & 0 & 0 \\ 0 & n_x & -n_y \\ 0 & n_y & n_x \end{bmatrix} \tag{7.12}$$

$$\mathbf{F} \cdot \mathbf{n} = \mathbf{E}n_x + \mathbf{G}n_y = \mathbf{T}^{-1}\mathbf{E}(\mathbf{T}\mathbf{U}) \tag{7.13}$$

$$\tilde{\mathbf{F}} = \mathbf{T}^{-1}\tilde{\mathbf{E}}(\mathbf{Q}) \tag{7.14}$$

$$\mathbf{Q} = \mathbf{T}\mathbf{U} = \begin{pmatrix} h \\ q_x n_x + q_y n_y \\ -q_x n_y + q_y n_x \end{pmatrix}, \quad \mathbf{E}(\mathbf{Q}) = \begin{pmatrix} q_x n_x + q_y n_y \\ \dfrac{\left(q_x n_x + q_y n_y\right)^2}{h} + \dfrac{gh^2}{2} \\ \dfrac{(q_x n_x + q_y n_y)(-q_x n_y + q_y n_x)}{h} \end{pmatrix} \tag{7.15}$$

$$\tilde{\mathbf{E}}^{HLL} = \begin{cases} \mathbf{E}(\mathbf{Q}_L) & \text{if} & 0 \le S_L \\[2mm] \dfrac{S_R \mathbf{E}(\mathbf{Q}_L) - S_L \mathbf{E}(\mathbf{Q}_R) + S_L S_R(\mathbf{Q}_R - \mathbf{Q}_L)}{S_R - S_L} & \text{if} & S_L < 0 < S_R \\[2mm] \mathbf{E}(\mathbf{Q}_R) & \text{if} & 0 \ge S_R \end{cases} \tag{7.16}$$

$$\tilde{\mathbf{E}}^{HLLC} = \begin{cases} \mathbf{E}(\mathbf{Q}_L) & \text{if} & S_L \ge 0 \\ \mathbf{E}(\mathbf{Q}_L) + S_L(\mathbf{Q}_{*L} - \mathbf{Q}_L) & \text{if} & S_L < 0 \le S_* \\ \mathbf{E}(\mathbf{Q}_R) + S_R(\mathbf{Q}_{*R} - \mathbf{Q}_R) & \text{if} & S_* < 0 < S_R \\ \mathbf{E}(\mathbf{Q}_R) & \text{if} & S_R \le 0 \end{cases} \tag{7.17}$$

$$S_L = \min\left(u_{nL} - \sqrt{gh_L}, u_n^* - \sqrt{gh^*}\right) \tag{7.18}$$

$$S_R = \max\left(u_{nR} + \sqrt{gh_R}, u_n^* + \sqrt{gh^*}\right) \tag{7.19}$$

$$S_* = \frac{S_L h_R(u_{nR} - S_R) - S_R h_L(u_{nL} - S_L)}{h_R(u_{nR} - S_R) - h_L(u_{nL} - S_L)} \tag{7.20}$$

$$u_n^* = \frac{1}{2}(u_{nL} + u_{nR}) + \sqrt{gh_L} - \sqrt{gh_R} \tag{7.21}$$

$$\sqrt{gh^*} = \frac{1}{2}\left(\sqrt{gh_L} + \sqrt{gh_R}\right) + \frac{1}{4}(u_{nL} - u_{nR}) \tag{7.22}$$

$$\mathbf{Q}_{*(L,R)} = h_{(L,R)}\left(\frac{S_{(L,R)} - u_{n(L,R)}}{S_{(L,R)} - S_*}\right) \begin{bmatrix} 1 \\ S_* \\ u_{t(L,R)} \end{bmatrix} \tag{7.23}$$

The Roe flux for 2D shallow water flow equations (Wang and Liu, 2000) is given by Equation (7.24). The parameters of the Roe flux are given by Equations (7.25) to (7.28). The HLLC numerical flux is adopted in the following tests unless specified otherwise.

$$\tilde{\mathbf{F}}^{Roe} = \frac{1}{2}(\mathbf{F}_L + \mathbf{F}_R) \cdot \mathbf{n} - \frac{1}{2}\sum_{i=1}^{3}\tilde{\alpha}_i |\tilde{\lambda}_i| \tilde{\mathbf{K}}_i \tag{7.24}$$

$$\left. \begin{aligned} \tilde{\alpha}_1 &= \frac{\Delta h}{2} - \frac{1}{2\tilde{c}}\left[\Delta q_x n_x + \Delta q_y n_y - (\tilde{u}n_x + \tilde{v}n_y)\Delta h\right] \\ \tilde{\alpha}_2 &= \frac{1}{\tilde{c}}[(\Delta q_y - \tilde{v}\Delta h)n_x - (\Delta q_x - \tilde{u}\Delta h)n_y] \\ \tilde{\alpha}_3 &= \frac{\Delta h}{2} + \frac{1}{2\tilde{c}}[\Delta q_x n_x + \Delta q_y n_y - (\tilde{u}n_x + \tilde{v}n_y)\Delta h] \end{aligned} \right\} \tag{7.25}$$

$$\left. \begin{aligned} \tilde{\lambda}_1 &= \tilde{u}n_x + \tilde{v}n_y - \tilde{c} \\ \tilde{\lambda}_2 &= \tilde{u}n_x + \tilde{v}n_y \\ \tilde{\lambda}_3 &= \tilde{u}n_x + \tilde{v}n_y + \tilde{c} \end{aligned} \right\} \tag{7.26}$$

$$\tilde{\mathbf{K}}_1 = \begin{bmatrix} 1 \\ \tilde{u} - \tilde{c}n_x \\ \tilde{v} - \tilde{c}n_x \end{bmatrix}, \quad \tilde{\mathbf{K}}_2 = \begin{bmatrix} 0 \\ -\tilde{c}n_y \\ \tilde{c}n_x \end{bmatrix}, \quad \tilde{\mathbf{K}}_3 = \begin{bmatrix} 1 \\ \tilde{u} + \tilde{c}n_x \\ \tilde{v} + \tilde{c}n_x \end{bmatrix} \tag{7.27}$$

$$\left. \begin{aligned} \tilde{u} &= \frac{u_R\sqrt{h_R} + u_L\sqrt{h_L}}{\left(\sqrt{h_L} + \sqrt{h_R}\right)} \\ \tilde{v} &= \frac{v_R\sqrt{h_R} + v_L\sqrt{h_L}}{\left(\sqrt{h_L} + \sqrt{h_R}\right)} \\ \tilde{c} &= \sqrt{\frac{g}{2}(h_L + h_R)} \end{aligned} \right\} \tag{7.28}$$

## 7.3   Dry bed treatment

For dry bed problems, the numerical flux is evaluated through a two-wave HLL flux function, and the wave speed for right-hand dry bed and left-hand dry bed boundaries are given, respectively, by Equations

(7.29) and (7.30). Dry bed treatment similar to the one-dimensional case is adopted, a small depth is assigned at the dry nodes or a small depth criterion to track wet/dry front with zero depth at the dry nodes is used. Numerical tests show that for horizontal beds or channel beds with small bed variations, these two dry bed treatments provide similar results. However, for large variations in bed geometry, the dry bed treatment with zero depth and small depth to track the wet/dry front give more accurate results.

$$\left. \begin{aligned} S_L &= u_{nL} - \sqrt{gh_L} \\ S_R &= u_{nL} + 2\sqrt{gh_L} \end{aligned} \right\} \tag{7.29}$$

$$\left. \begin{aligned} S_L &= u_{nR} - 2\sqrt{gh_R} \\ S_R &= u_{nR} + \sqrt{gh_R} \end{aligned} \right\} \tag{7.30}$$

## 7.4   Numerical tests

In this section, numerical tests in channels with horizontal beds are simulated. The bed elevation term is neglected and only the friction term is considered in the source term. Several tests are conducted to verify the numerical scheme.

### 7.4.1   Oblique hydraulic jump

In this test, numerical simulation of an oblique hydraulic jump in a channel with a horizontal bed is presented. The plan view of the channel and the computational mesh are shown in Figure 7.1. An oblique hydraulic jump forms inside the channel as the supercritical flow is deflected by the converging wall. Water depth ($h$) of 1.0 m and longitudinal velocity ($u$) of 8.57 m/s, and transverse velocity ($v$) of 0 m/s are prescribed as inflow boundary conditions. No boundary condition is applied at the outflow boundary. Based on the analytical solution, the shock angle is 30° and the water depth is 1.5049 m after the shock. The computed water surface at a steady state is shown in Figure 7.2, and the water surface contours are shown in Figure 7.3. Comparison between simulated results with different flux functions and analytical solutions along the solid line, shown in Figure 7.1, is presented in Figure 7.4. The computed solutions are in good agreement with the analytical solutions. The results with HLL, Roe, and HLLC fluxes are similar.

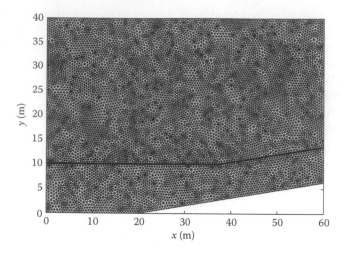

*Figure 7.1* Computational domain and mesh for the oblique hydraulic jump test.

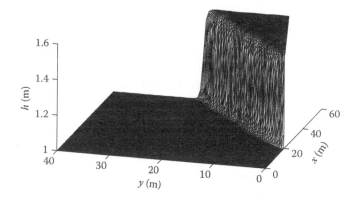

*Figure 7.2* Computed water surface profile at a steady state for the oblique hydraulic jump test.

### 7.4.2   Shock wave in a channel contraction

The DG scheme is used to simulate the steady supercritical shock wave in a channel contraction. Ippen and Dawson (1951) provided an analytical solution to the shock wave in a channel contraction. The channel geometry and mesh used are shown in Figure 7.5. The channel width at the upstream end is 20 m, and the width at the downstream end is 10.548 m. The wall deflection angle is 12° and the length of contraction is 22.234 m. Water depth of 1 m, longitudinal velocity of 8.4566 m/s, and zero transverse

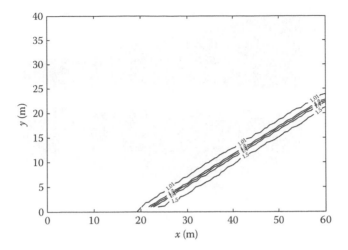

*Figure 7.3* Computed water depth contours for the oblique hydraulic jump test.

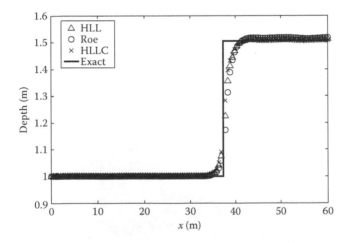

*Figure 7.4* Numerical and analytic solutions along the solid line for the oblique hydraulic jump test.

velocity are prescribed as inflow boundary conditions. No boundary condition is applied at the outflow end of the channel. The computed steady-state solutions for the water surface elevation and water depth contours are shown in Figures 7.6 and 7.7, respectively. Numerical and analytical solutions along the dash and solid lines (shown in Figure 7.5) are compared in Figures 7.8 and 7.9, respectively. The results show that different

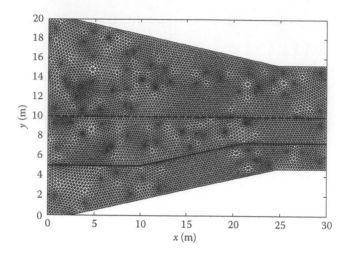

*Figure 7.5* Computational domain and mesh for the shock wave test.

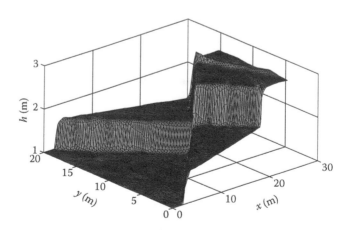

*Figure 7.6* Computed water surface profile in 3D for the shock wave test.

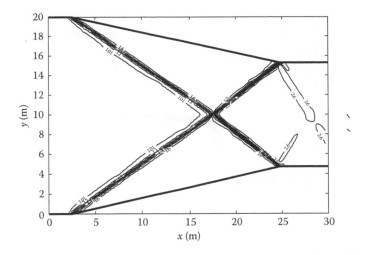

***Figure 7.7*** Computed water depth contours for the shock wave test.

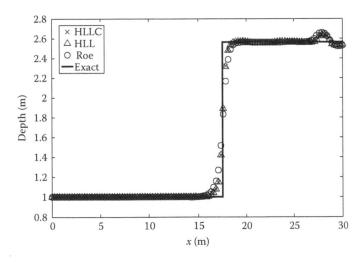

***Figure 7.8*** Numerical and analytic solutions along the dashed line for the shock wave test.

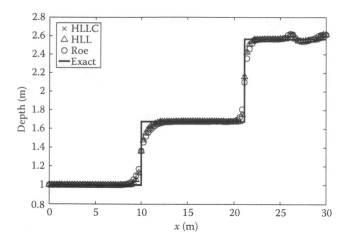

**Figure 7.9** Numerical and analytic solutions along the solid line for the shock wave test.

numerical flux functions provide similar accuracy. The shock waves are captured accurately by the numerical scheme. Some oscillations are observed after the shock.

### 7.4.3  Circular dam break

Two different cases of a circular dam break are simulated in this test. A circular dam with a radius of 11 m is located in the center of a 50 m × 50 m solid wall container. For the outflow dam-break case, the water depth is initially 10 m inside the dam and 1 m outside. For the inflow dam-break case, the water depth is initially 1 m inside the dam and 10 m outside. The circular dam is removed instantaneously. The computed water surface and the water depth contours for the outflow dam-break case at 0.8 s are shown in Figures 7.10 and 7.11, respectively. The computed water surface and the water depth contours for the inflow dam-break case at 2.0 s are presented in Figures 7.12 and 7.13, respectively. The symmetric shock waves are captured accurately by the numerical scheme.

### 7.4.4  Partial dam break

The channel configuration and mesh for the partial dam-break case are shown in Figure 7.14. The horizontal channel has a dam in the middle with a breach that is 75 m wide. The partial dam-break tests with the wet bed and dry bed downstream of the dam are simulated in this test. The

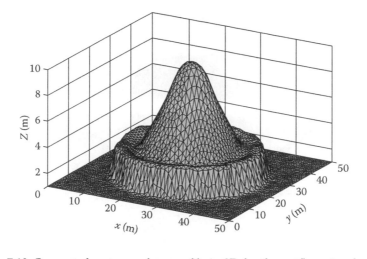

*Figure 7.10* Computed water surface profile in 3D for the outflow circular dam-break test at 0.8 s.

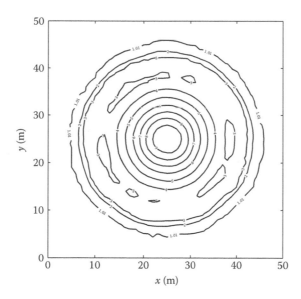

*Figure 7.11* Computed water depth contours for the outflow circular dam-break test at 0.8 s.

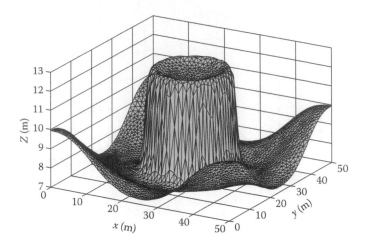

*Figure 7.12* Computed water surface profile in 3D for the inflow circular dam-break test at 2 s.

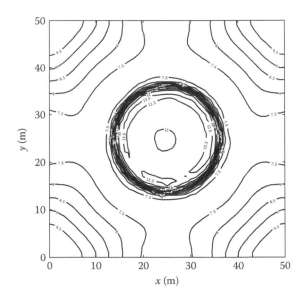

*Figure 7.13* Computed water depth contour for the inflow circular dam-break test at 2 s.

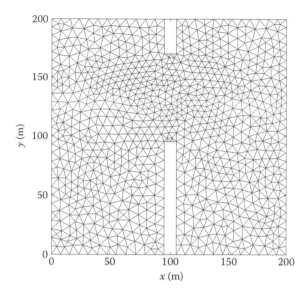

*Figure 7.14* Computational domain for the partial dam-break test.

water depth upstream of the dam is 10 m, and the downstream water depth is 5 m and zero for the wet bed and dry bed cases, respectively. The dam break is assumed to take place instantaneously. For the wet bed case, the water surface and the water depth contours in 3D view at 7 s after the dam break are shown in Figures 7.15 and 7.16, respectively. For the dry bed

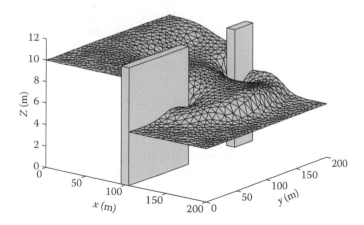

*Figure 7.15* Water surface profile in 3D for the partial dam-break test with a wet bed downstream at 7 s.

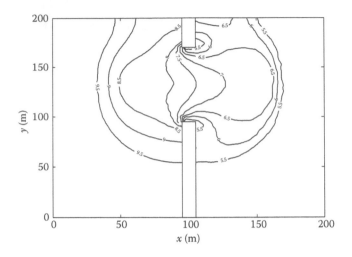

**Figure 7.16** Water depth contours for the partial dam-break test with a wet bed downstream at 7 s.

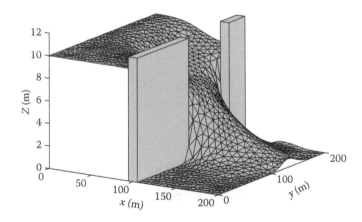

**Figure 7.17** Water surface profile in 3D for the partial dam-break test with a dry bed downstream at 7 s.

case, the water surface and the water depth contours at 7 s after the dam break are presented in Figures 7.17 and 7.18, respectively.

## 7.4.5 Dam-break flows in channels with bends

Experiments related to dam-break flows in channels with sharp bends were conducted in the Civil Engineering Department Laboratory,

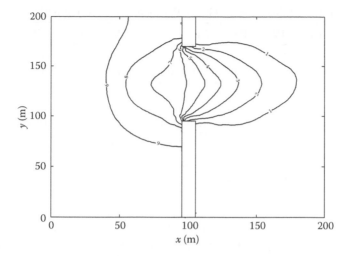

**Figure 7.18** Water depth contours for the partial dam-break test with a dry bed downstream at 7 s.

Université Catholique de Louvain (UCL, Belgium) and used by the Concerted Action on Dambreak Modeling (CADAM) project (Frazão et al., 1998). These experiments are simulated using the DG method. The plan views of the channels with a 45° bend and a 90° bend are shown in Figures 7.19 and 7.29, respectively, along with the location of the gauge points. The coordinates of the gauge points are given in Tables 7.1 and 7.2. There is a 244 cm × 239 cm reservoir at the upstream end of the channel, and the rectangular channel is 49.5 cm wide. For the 45° bend case, the gauges are labeled as P1 through P9, and for the 90° bend case the gauges are denoted by G1 through G6. The Manning roughness coefficient of

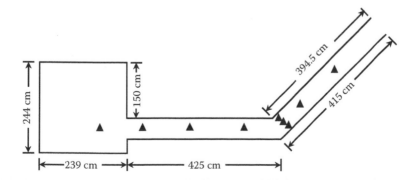

**Figure 7.19** Plan view of the channel with a 45° bend.

*Table 7.1* Coordinates for the Gauge Points in the Channel with a 45° Bend

| Gauge | P1 | P2 | P3 | P4 | P5 | P6 | P7 | P8 | P9 |
|-------|------|------|------|------|------|------|------|------|------|
| x (m) | 1.59 | 2.74 | 4.24 | 5.74 | 6.74 | 6.65 | 6.56 | 7.07 | 8.13 |
| y (m) | 0.69 | 0.69 | 0.69 | 0.69 | 0.72 | 0.80 | 0.89 | 1.22 | 2.28 |

*Table 7.2* Coordinates for the Gauge Points in the Channel with a 90° Bend

| Gauge | G1 | G2 | G3 | G4 | G5 | G6 |
|-------|------|------|------|------|------|------|
| x (m) | 1.19 | 2.74 | 4.24 | 5.74 | 6.56 | 6.56 |
| y (m) | 1.20 | 0.69 | 0.69 | 0.69 | 1.51 | 3.01 |

$0.0095$ s/m$^{1/3}$ is used for the channel bed and $0.0195$ s/m$^{1/3}$ for the walls of the channel. The gate at the outlet of the reservoir represents the dam. Instantaneous failure of the dam is assumed. The water depth in the reservoir is initially 0.25 m for the 45° bend case, and 0.2 m for the 90° bend case. For both cases, the bed downstream is dry. For the 45° bend case, the computational domain is triangulated with 2316 elements, while for the 90° bend case, 8546 elements are used. The computed and measured hydrographs for P1 through P9 are compared in Figures 7.20 to 7.28, respectively, and for G1 through G6 in Figures 7.30 to 7.35, respectively. The model performs well in simulating dam-break wave and reflections generated from bends.

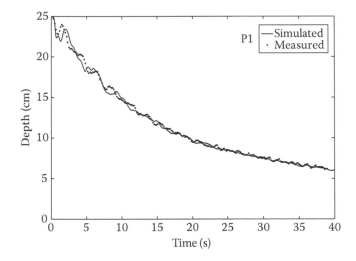

*Figure 7.20* Simulated and measured hydrographs at P1 for the dam-break test in the channel with a 45° bend.

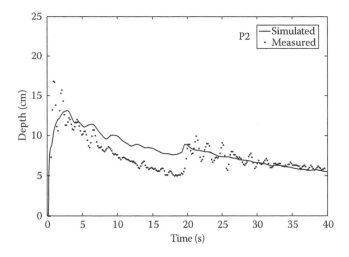

*Figure 7.21* Simulated and measured hydrographs at P2 for the dam-break test in the channel with a 45° bend.

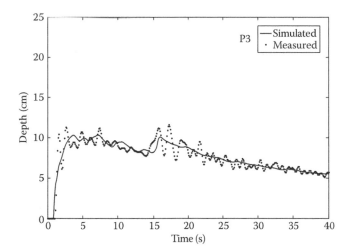

*Figure 7.22* Simulated and measured hydrographs at P3 for the dam-break test in the channel with a 45° bend.

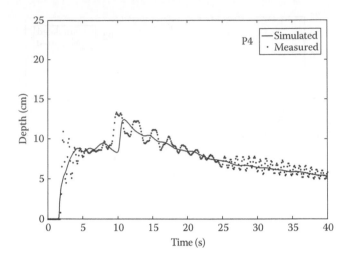

**Figure 7.23** Simulated and measured hydrographs at P4 for the dam-break test in the channel with a 45° bend.

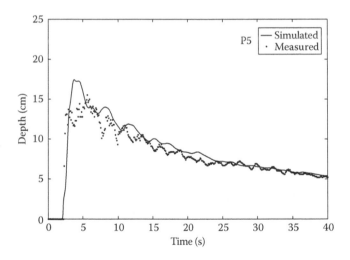

**Figure 7.24** Simulated and measured hydrographs at P5 for the dam-break test in the channel with a 45° bend.

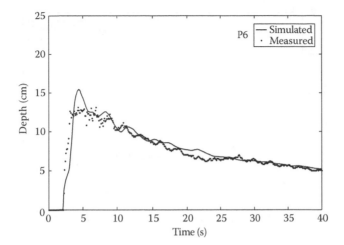

*Figure 7.25* Simulated and measured hydrographs at P6 for the dam-break test in the channel with a 45° bend.

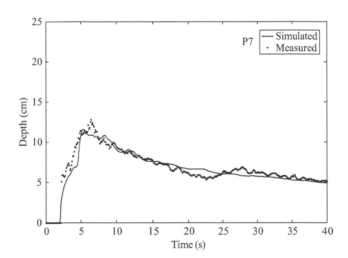

*Figure 7.26* Simulated and measured hydrographs at P7 for the dam-break test in the channel with a 45° bend.

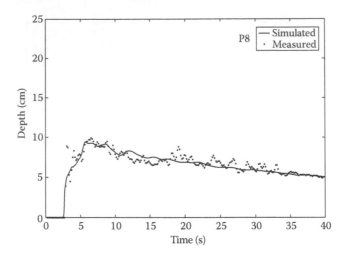

*Figure 7.27* Simulated and measured hydrographs at P8 for the dam-break test in the channel with a 45° bend.

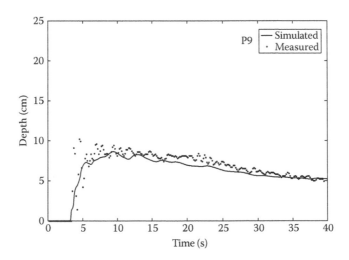

*Figure 7.28* Simulated and measured hydrographs at P9 for the dam-break test in the channel with a 45° bend.

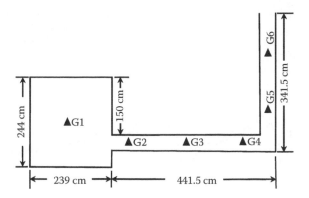

*Figure 7.29* Plan view of the channel with a 90° bend.

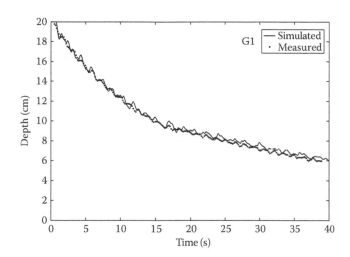

*Figure 7.30* Simulated and measured hydrographs at G1 for the dam-break test in the channel with a 90° bend.

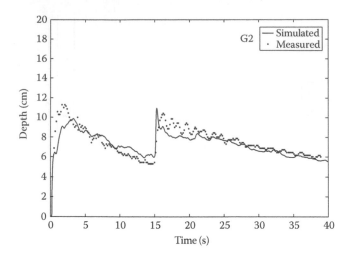

**Figure 7.31** Simulated and measured hydrographs at G2 for the dam-break test in the channel with a 90° bend.

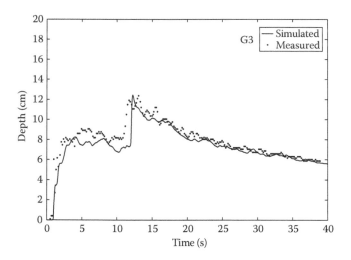

**Figure 7.32** Simulated and measured hydrographs at G3 for the dam-break test in the channel with a 90° bend.

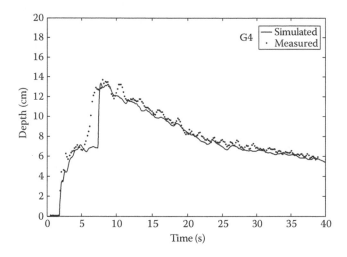

**Figure 7.33** Simulated and measured hydrographs at G4 for the dam-break test in the channel with a 90° bend.

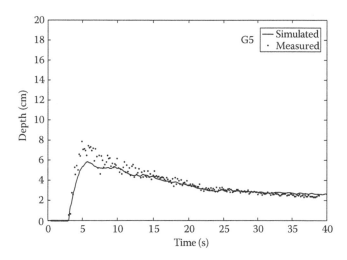

**Figure 7.34** Simulated and measured hydrographs at G5 for the dam-break test in the channel with a 90° bend.

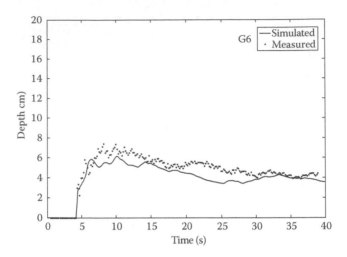

*Figure 7.35* Simulated and measured hydrographs at G6 for the dam-break test in the channel with a 90° bend.

# References

Eskilsson, C., and Sherwin, S. J. (2004). A triangular spectral/hp discontinuous Galerkin method for modelling 2D shallow water equations. *International Journal for Numerical Methods in Fluids*, 45(6), 605–623.

Fraccarollo, L., and Toro, E. F. (1995). Experimental and numerical assessment of the shallow water model for two-dimensional dam-break type problems. *Journal of Hydraulic Research*, 33(6), 843–846.

Frazão, S. S., Sillen, X., and Zech, Y. (1998). Dam-break flow through sharp bends, physical model and 2D Boltzmann model validation. Proceedings of the 1st CADAM Meeting, Wallingford, UK.

Ippen, A. T., and Dawson, J. H. (1951). Design of channel contraction. *Transactions of the American Society of Civil Engineers*, 116, 326–346.

Lai, W., and Khan, A. A. (2012). A discontinuous Galerkin method for two-dimensional shallow water flows. *International Journal for Numerical Methods in Fluids*, 70(8), 939–960.

Wang, J. W., and Liu, R. X. (2000). A comparative study of finite volume methods on unstructured meshes for simulation of 2D shallow water wave problems. *Mathematics and Computers in Simulation*, 53(3), 171–184.

*chapter eight*

---

# Two-dimensional shallow water flow in channels with bed variations

To model two-dimensional (2D) flow in channels with bed variation, the depth and bed level variations must be accounted for. To maintain the well balance property (that is to avoid bed level change generated flows) of the numerical scheme, a form of the two-dimensional shallow water flow equations that is suitable for channels with bed variations is presented. The application of the slope limiter and flux approximation for this form of equations is described. The resulting model is tested using laboratory and natural channel flows.

## 8.1 Two-dimensional shallow water flow equations for natural channels

Besides the form of the two-dimensional shallow water flow equations presented in Chapter 7, these equations can also be written in the form given by Equation (8.1). In this form, the bed slope and hydrostatic force terms are combined into a water surface gradient term. The governing equations can be written in conservation form as shown in Equation (8.2). The conserved variable $\mathbf{U}$, source term $\mathbf{S}$, and flux terms $\mathbf{E}$ and $\mathbf{G}$ are given by Equation (8.3). For this form of the equation, the vectors $\mathbf{Q}$ and $\mathbf{E}(\mathbf{Q})$ are given by Equation (8.4).

$$\left. \begin{aligned}
\frac{\partial h}{\partial t} + \frac{\partial q_x}{\partial x} + \frac{\partial q_y}{\partial y} &= 0 \\[2mm]
\frac{\partial q_x}{\partial t} + \frac{\partial q_x^2/h}{\partial x} + \frac{\partial q_x q_y/h}{\partial y} &= -gh\frac{\partial Z}{\partial x} - ghS_{fx} \\[2mm]
\frac{\partial q_y}{\partial t} + \frac{\partial q_x q_y/h}{\partial x} + \frac{\partial q_y^2/h}{\partial y} &= -gh\frac{\partial Z}{\partial y} - ghS_{fy}
\end{aligned} \right\} \qquad (8.1)$$

$$\frac{\partial \mathbf{U}}{\partial t} + \nabla \cdot \mathbf{F}(\mathbf{U}) = \frac{\partial \mathbf{U}}{\partial t} + \frac{\partial \mathbf{E}(\mathbf{U})}{\partial x} + \frac{\partial \mathbf{G}(\mathbf{U})}{\partial x} = \mathbf{S}(\mathbf{U}) \qquad (8.2)$$

$$\mathbf{U} = \begin{pmatrix} h \\ q_x \\ q_y \end{pmatrix}, \quad \mathbf{S} = \begin{pmatrix} 0 \\ -gh\dfrac{\partial Z}{\partial x} - ghS_{fx} \\ -gh\dfrac{\partial Z}{\partial y} - gghS_{fy} \end{pmatrix},$$

$$\mathbf{E(U)} = \begin{pmatrix} q_x \\ q_x^2/h \\ q_xq_y/h \end{pmatrix}, \quad \mathbf{G(U)} = \begin{pmatrix} q_y \\ q_xq_y/h \\ q_y^2/h \end{pmatrix}$$

(8.3)

$$\mathbf{Q = TU} = \begin{pmatrix} h \\ q_xn_x + q_yn_y \\ -q_xn_y + q_yn_x \end{pmatrix}, \quad \mathbf{E(Q)} = \begin{pmatrix} q_xn_x + q_yn_y \\ \dfrac{\left(q_xn_x + q_yn_y\right)^2}{h} \\ \dfrac{(q_xn_x + q_yn_y)(-q_xn_y + q_yn_x)}{h} \end{pmatrix}$$

(8.4)

## 8.2   Numerical flux and source term treatment

To use the HLLC (Harten-Lax-van Leer contact) flux with Equation (8.1), the wave speeds given in Chapter 7 by Equations (7.18) to (7.20) should be used. Using the notation for the elements shown in Figure 6.2 (see Chapter 6), the calculation of the water level slopes in the source vector can be determined with the Green theorem (Ying et al., 2009) as given by Equations (8.5) and (8.6). The water surface slope terms given by Equations (8.5) and (8.6) are treated as constants in evaluating integrals.

$$\Omega_0\frac{\partial Z}{\partial x} = \oint_{\Gamma_0} Z\,dy = \sum_{k=1}^{3} Z_{0k}\Delta y_k$$
$$= Z_{01}(y_{n2} - y_{n1}) + Z_{02}(y_{n3} - y_{n2}) + Z_{03}(y_{n1} - y_{n3})$$

(8.5)

$$-\Omega_0\frac{\partial Z}{\partial y} = \oint_{\Gamma_0} Z\,dx = \sum_{k=1}^{3} Z_{0k}\Delta x_k$$
$$= Z_{01}(x_{n2} - x_{n1}) + Z_{02}(x_{n3} - x_{n2}) + Z_{03}(x_{n1} - x_{n3})$$

(8.6)

In these equations, $Z_{0k}$ is the water level at the boundary of elements 0 and $k$, and $\Omega_0$ is the area of element 0. The water level, $Z_{0k}$, can be determined using the water surface elevations of elements 0 and $k$. First, the average water surface elevations at the center of all elements are determined and then the water surface elevations at the boundaries are interpolated using inverse distance weighting. The discretization guarantees that if the water surface in the main element and the surrounding element is the same, then there will be no bed-topography generated unphysical flows. Numerical results show that this treatment of the source term is accurate.

The well-balanced property in the wet domain is satisfied with Equation (8.1). The still water with a partially wet domain can be easily achieved by setting the source term **S** to zero in both partially wet elements and dry elements. In addition, the slope limiter is not applied in elements with zero velocities at all three nodes. Although the source term in partially wet elements is forced to be zero, numerical tests in dam-break flows show that the flood waves are still accurately modeled.

## 8.3    Numerical tests in channels with irregular beds

In case of irregular beds, the bed elevation and the pressure force terms are combined to obtain the water surface elevation term. The water surface elevation term is used as part of the source term. Several tests are shown next to verify the formulation in the discontinuous Galerkin (DG) scheme.

### 8.3.1    Dam-break flow in a channel with a triangular bump

In this test, the numerical scheme is used to model dam-break flow over a triangular bump (Hiver, 2000) as proposed by the Concerted Action on Dambreak Modeling (CADAM) project. Numerical results are compared with the experimental data. The initial condition and the bed topography are shown in Figure 8.1. The rectangular channel is 38 m long, 0.75 m wide

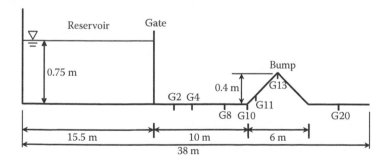

*Figure 8.1* Geometry and experimental set up of the channel with a triangular bump.

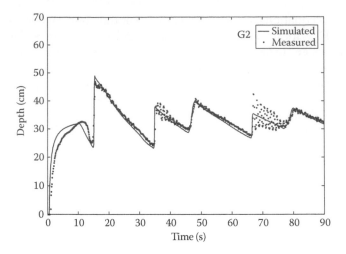

*Figure 8.2* Simulated and measured hydrographs at G2 for the dam-break test in the channel with a triangular bump.

with a gate located at 15.5 m from the upstream end. The symmetric triangular bump (6 m long, 0.4 m high) is situated at 13 m downstream of the gate. The water depth upstream of the gate is 0.75 m with a dry bed downstream. The suggested values of the Manning roughness coefficients of 0.0125 and 0.011 for the bed and the walls, respectively, are used. A free outflow boundary condition (i.e., no boundary condition) is applied at the outflow end. The computational domain is triangulated with 4352 elements. The dry bed depth of 0.001 m and time step of 0.006 s are used. Simulated and measured hydrographs at 90 seconds after the dam removal at gauge points G2, G4, G8, G10, G11, G13, and G20 are shown in Figures 8.2 to 8.8, respectively. The number of the gauge point denotes distance from the gate, for example, G2 is located 2 m downstream of the gate. The simulated results are in good agreement with the measured data. The flood wave arrival time and water depth are well predicted at all gauge points. The wetting and drying effect at the critical point G13, which is located at the vertex of the bump, is modeled correctly. The difference between simulated and measured results at the last point G20 mostly comes from the uncertainty of the actual outflow boundary condition.

## 8.3.2   Oscillating flow in a parabolic bowl

Flow in a parabolic bowl with axisymmetric oscillating free surface is simulated to test wetting and drying capabilities as well as mass and

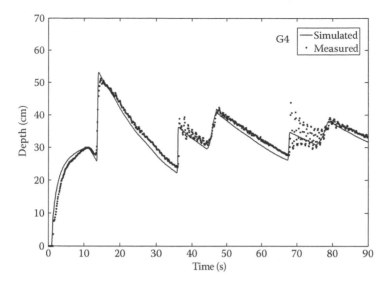

**Figure 8.3** Simulated and measured hydrographs at G4 for the dam-break test in the channel with a triangular bump.

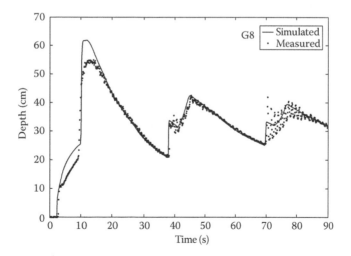

**Figure 8.4** Simulated and measured hydrographs at G8 for the dam-break test in the channel with a triangular bump.

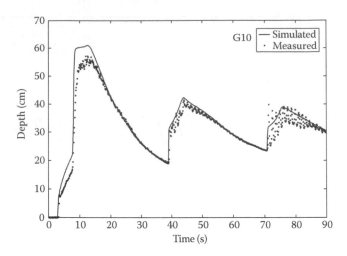

***Figure 8.5*** Simulated and measured hydrographs at G10 for the dam-break test in the channel with a triangular bump.

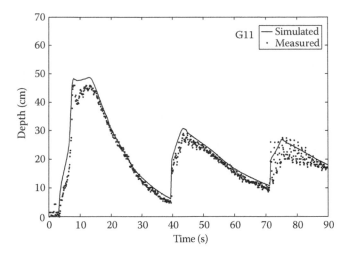

***Figure 8.6*** Simulated and measured hydrographs at G11 for the dam-break test in the channel with a triangular bump.

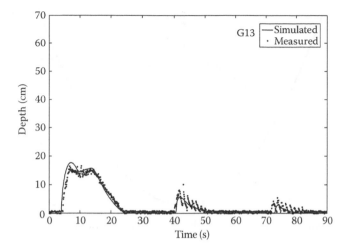

**Figure 8.7** Simulated and measured hydrographs at G13 for the dam-break test in the channel with a triangular bump.

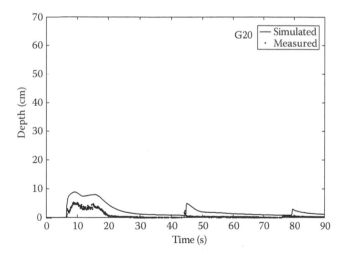

**Figure 8.8** Simulated and measured hydrographs at G20 for the dam-break test in the channel with a triangular bump.

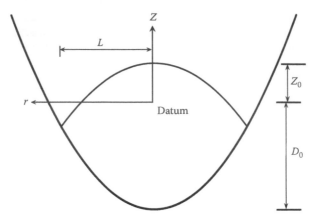

**Figure 8.9** Bottom profile of the parabolic bowl and initial water surface profile.

momentum conservation properties of the model. The bed profile of the parabolic bowl is shown in Figure 8.9 and is defined by Equation (8.7). In the equation, $D_o$ is the distance from the datum to the bottom of the bed at the center of the parabola and $L$ is the distance from the center to the shoreline. For the frictionless case, the analytic solution for the water surface elevation and the velocity in the wet region are given by Thacker (1981) and are shown in Equations (8.8) and (8.9), respectively. The parameters $A$ and $\omega$ are given by Equations (8.10) and (8.11), respectively, where $Z_o$ is the water surface elevation at the center of the bowl. The wet region in which the water depth is greater than zero is given by Equation (8.12).

$$z_b = D_o\left(\frac{r^2}{L^2}-1\right), \quad r = \sqrt{x^2+y^2} \tag{8.7}$$

$$Z(x,y,t) = D_o\left\{\frac{\sqrt{(1-A^2)}}{1-A\cos(\omega t)}-1-\frac{r^2}{L^2}\left[\frac{(1-A^2)}{(1-A\cos(\omega t))^2}-1\right]\right\} \tag{8.8}$$

$$(u(x,y,t), v(x,y,t)) = \frac{1}{2}\left[\frac{\omega A\sin(\omega t)}{1-A\cos(\omega t)}\right](x,y) \tag{8.9}$$

$$A = \frac{(D_o+Z_o)^2-D_o^2}{(D_o+Z_o)^2+D_o^2} \tag{8.10}$$

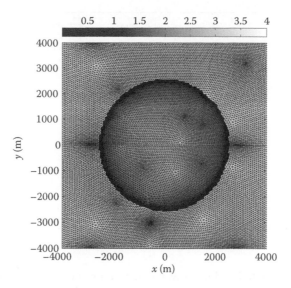

*Figure 8.10* Computational domain and initial water depth in the parabolic bowl.

$$\omega = \frac{2\pi}{T} = \frac{\sqrt{8gD_o}}{L} \qquad (8.11)$$

$$r^2 < \frac{L^2(1 - A\cos(\omega t))}{\sqrt{1 - A^2}} \qquad (8.12)$$

In this simulation, $D_o = 3$ m, $Z_o = 1$ m, and $L = 3000$ m are used, which gives the oscillation period $T = 2457$ s. The computational domain is shown in Figure 8.10 along with the initial water surface level. The domain is triangulated with 27648 elements and 14033 nodes. Numerical results for the water surface and flow rate at different times along the line $y = 0$ are shown in Figures 8.11 to 8.18. Comparison of numerical results and exact solutions shows the scheme is capable of the modeling wetting and drying process, and can conserve mass and momentum.

### 8.3.3   Dam-break flow with three bumps in the downstream channel

The model is used to simulate dam-break flow with three bumps in the downstream channel. The channel is 75 m long and 30 m wide with a

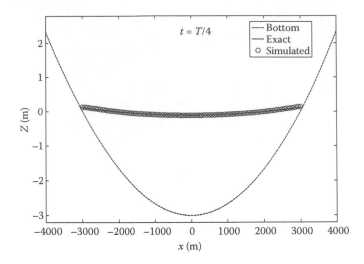

***Figure 8.11*** Computed and exact water surface profiles in the parabolic bowl at $t = T/4$.

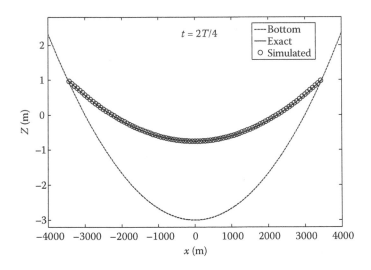

***Figure 8.12*** Computed and exact water surface profiles in the parabolic bowl at $t = 2T/4$.

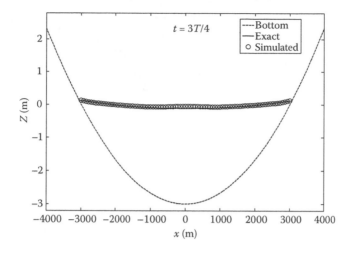

***Figure 8.13*** Computed and exact water surface profiles in the parabolic bowl at $t = 3T/4$.

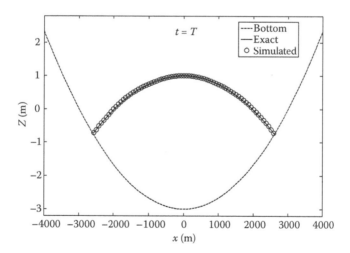

***Figure 8.14*** Computed and exact water surface profiles in the parabolic bowl at $t = T$.

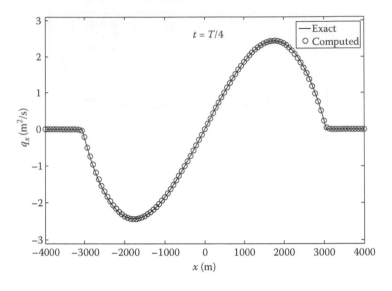

**Figure 8.15** Computed and exact flow rates in the parabolic bowl at $t = T/4$.

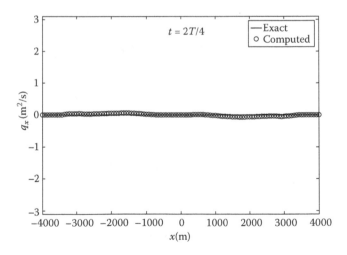

**Figure 8.16** Computed and exact flow rates in the parabolic bowl at $t = 2T/4$.

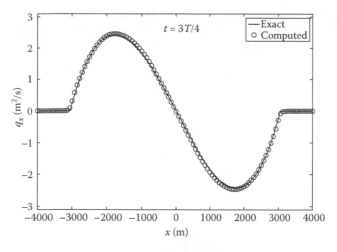

*Figure 8.17* Computed and exact flow rates in the parabolic bowl at $t = 3T/4$.

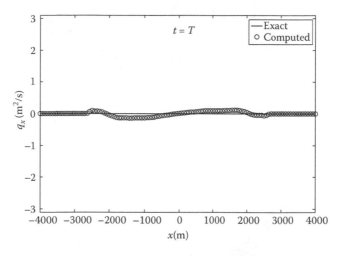

*Figure 8.18* Computed and exact flow rates in the parabolic bowl at $t = T$.

closed wall, and the channel bed is defined by Equation (8.13). A dam is located at $x = 16$ m and retains water to a depth of 1.875 m upstream of the dam and the bed downstream is dry. The bed topography and the initial water surface are shown in Figure 8.19. The dam is removed instantaneously and the water flow afterward is simulated. The dry bed depth criterion of 0.001 m, time step of 0.01 s, and the Manning roughness coefficient of $0.018 \text{ s/m}^{1/3}$ are used for the test. Simulated water surface profiles at different times are shown in Figures 8.20 to 8.24. The results show that the scheme is able to model flood propagation with wetting and drying effects in a channel with a highly irregular bed.

$$z_b(x,y) = \max \begin{bmatrix} 0, \quad 1-\frac{1}{8}\sqrt{(x-30)^2 + (y-6)^2}, \\ 1-\frac{1}{8}\sqrt{(x-30)^2 + (y-24)^2}, \\ 1-\frac{3}{10}\sqrt{(x-47.5)^2 + (y-15)^2} \end{bmatrix} \tag{8.13}$$

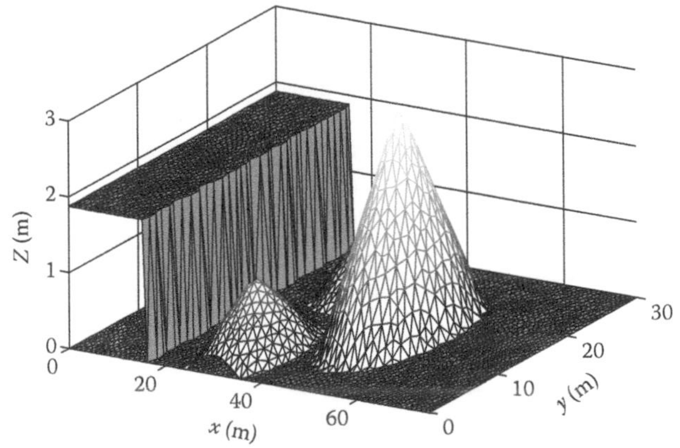

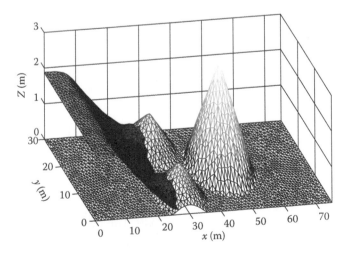

**Figure 8.19** Initial condition and bed topography of the channel with three bumps.

**Figure 8.20** Water surface profile for the dam-break flow with three bumps at 2 s.

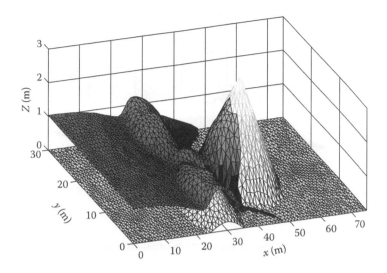

*Figure 8.21* Water surface profile for the dam-break flow with three bumps at 6 s.

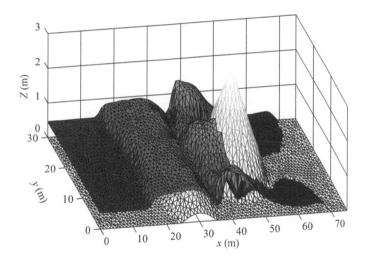

*Figure 8.22* Water surface profile for the dam-break flow with three bumps at 12 s.

## 8.3.4   The Toce River dam-break test

The dam-break flood in Toce River is simulated. The physical model of the river was constructed at the ENEL-HYDRO (Ente Nazionale per l'Energia Elettrica) Laboratory using a 1:100 scale of the Toce River valley reach, and used in the CADAM project (Frazão and Testa, 1999). The Manning roughness coefficient is taken to be 0.0162 s/m$^{1/3}$ in this test, and other modeling parameters such as topographic data and inflow

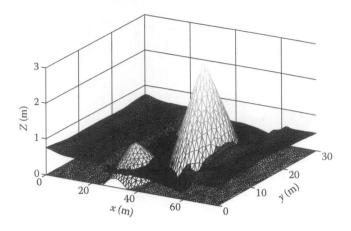

*Figure 8.23* Water surface profile for the dam-break flow with three bumps at 30 s.

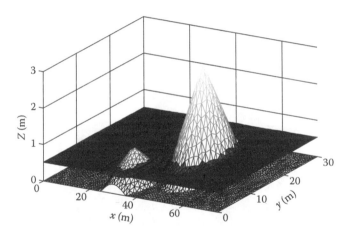

*Figure 8.24* Water surface profile for the dam-break flow with three bumps at 300 s.

hydrographs are provided by Electricité de France (EDF). The topography of the Toce River physical model is shown in Figure 8.25. It covered an area of 46 m × 10 m. The river was initially dry with a rectangular tank located upstream. Two different tests were conducted in the physical model, one without the overtopping and the other with overtopping of the reservoir in the downstream river reach. The corresponding inflow hydrographs (HY1 and HY2, respectively) are shown in Figure 8.26 for these two tests and used as inflow boundary conditions. For both tests,

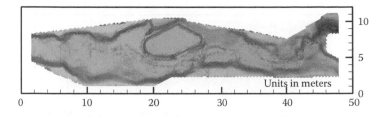

*Figure 8.25* Toce River bed topography.

critical flow boundary conditions are used at the inlet and outlet sections of the channel.

The water depth contours at $t = 25$ s, 40 s, and 60 s are shown in Figures 8.27 to 8.32 for both cases. The flood propagation along the river is well predicted. The comparisons between the computed and measured hydrographs at 5 gauge points are shown in Figures 8.33 and 8.34 for the two test cases. The positions of these gauges are shown in Table 8.1. The flood arrival time and water level are computed accurately.

## 8.3.5    The Paute River dam-break flood

This case is suitable to test the numerical scheme's ability to model dam-break flow in natural rivers with complex geometry, irregular bed topography,

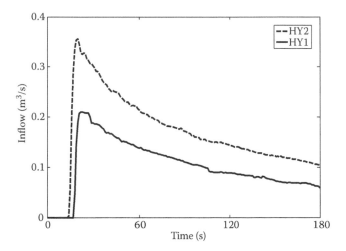

*Figure 8.26* Discharge from upstream tank for the Toce River test.

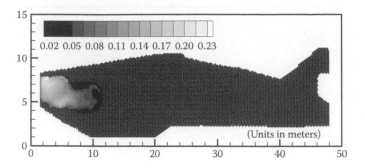

*Figure 8.27* Water depth in the Toce River at 25 s with HY1.

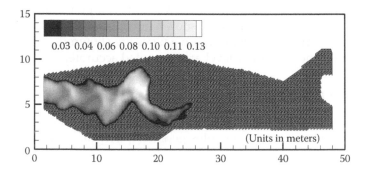

*Figure 8.28* Water depth in the Toce River at 40 s with HY1.

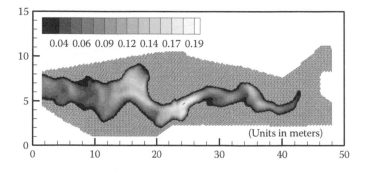

*Figure 8.29* Water depth in the Toce River at 60 s with HY1.

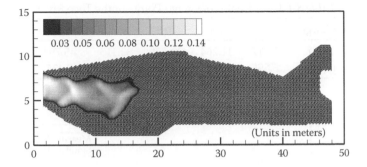

*Figure 8.30*  Water depth in the Toce River at 25 s with HY2.

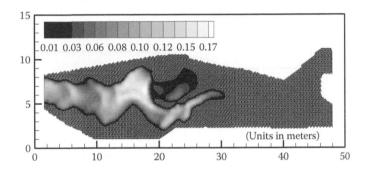

*Figure 8.31*  Water depth in the Toce River at 40 s with HY2.

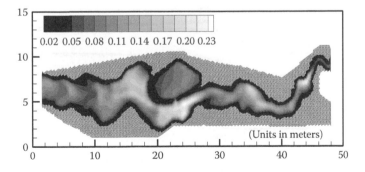

*Figure 8.32*  Water depth in the Toce River at 60 s with HY2.

*Table 8.1* Coordinates of Gauge Points in the Toce River

| Gauge | P1 | P5 | P13 | P21 | P26 |
|-------|------|------|------|------|------|
| $x$ (m) | 2.917 | 11.264 | 20.879 | 33.115 | 45.794 |
| $y$ (m) | 6.895 | 6.083 | 4.130 | 6.090 | 9.437 |

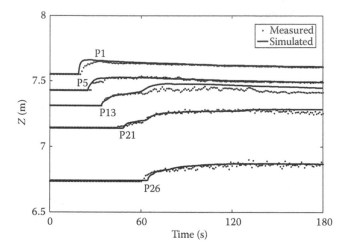

*Figure 8.33* Computed and measured stage–time hydrographs with HY1 for the Toce River test.

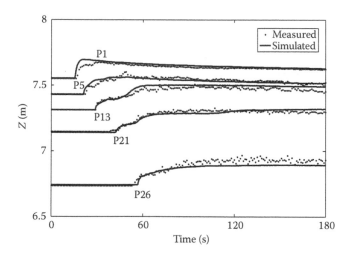

*Figure 8.34* Computed and measured stage–time hydrographs with HY2 for the Toce River test.

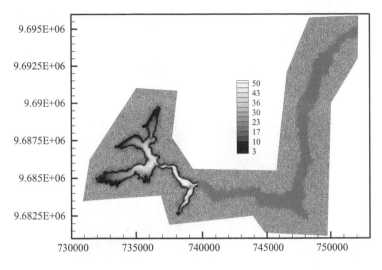

*Figure 8.35* Initial water depth, computation domain, and mesh in the Paute River.

and sharp bend. The dam is located in the Paute River, Ecuador. The river topography and triangulation data is available through BreZo (Sanders and Begnudelli, 2010). The computational domain with 74224 elements and initial water depth contours are shown in Figure 8.35 (units in meters). The dam is considered as a straight line between $(x, y)$ coordinates of (739602 m, 9684690 m) and (739616 m, 9684530 m), separating the upstream and downstream region. Initial water level upstream of the dam is 2362 m above sea level and the downstream channel is dry. A Manning roughness coefficient of 0.033 is used. The dam is assumed to fail instantaneously and completely for the simulation. The computed flood maps (water depth contours) at $t = 15$ min, 30 min, and 45 min are shown in Figures 8.36 to 8.38. Numerical results show the scheme is capable of simulating flood flow in natural rivers.

## 8.3.6   The Malpasset dam-break flood

The Malpasset Dam was located in the Reyran River Valley in France and failed in 1959 due to intense rainfall and a rapid increase of water level in the reservoir. The computational domain for the Malpasset dam-break flood is shown in Figure 8.39 along with 17 survey points (P1–P17 shown as circles) and 9 gauges (S6–S14 shown as crosses). The maximum water levels at the survey points were obtained after the dam failure, while the maximum water levels at the gauges were obtained through a physical model test with a scale of 1:400 built by Laboratoire National d'Hydraulique.

The bottom topography and measured data are provided by CADAM (Goutal, 1999). The computational domain covers an area of 17500 m × 9000 m.

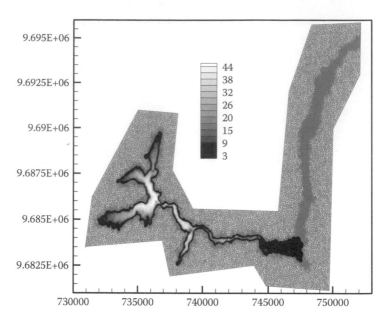

*Figure 8.36* Computed water depth in the Paute River at 15 min.

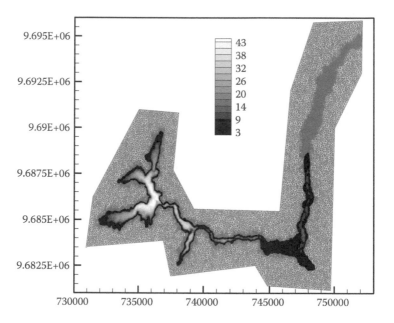

*Figure 8.37* Computed water depth in the Paute River at 30 min.

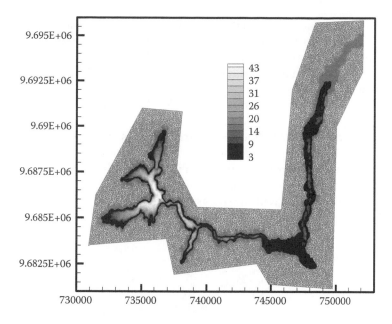

*Figure 8.38* Computed water depth in the Paute River at 45 min.

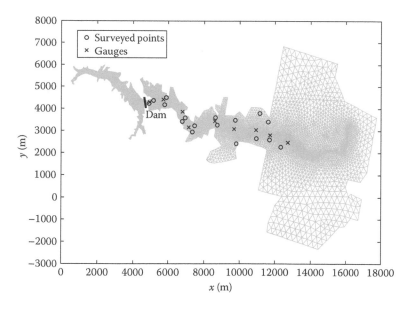

*Figure 8.39* Computational domain and survey points for the Malpasset test.

The dam is considered as a straight line between $(x, y)$ coordinates of (4701.18 m, 4143.41 m) and (4656.5 m, 4392.1 m), separating the upstream reservoir and the downstream dry bed. The initial water level in the reservoir is 100 m. The initial water discharge in the Reyran River is neglected. The Manning roughness coefficient of 0.029 is used.

The computed flood maps (water depth contours) at $t$ = 10 min, 20 min, and 30 min are shown in Figures 8.40 to 8.42 (units in meters). Computed and measured flood wave arrival times at gauges are shown in Figure 8.43. The computed and measured maximum water levels at gauges and surveyed points are presented in Figures 8.44 and 8.45, respectively. Simulated results for both flood arrival times and maximum water levels are in good agreement with the measured data.

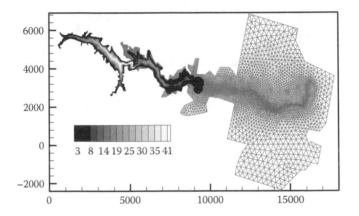

*Figure 8.40* Computed water depth for the Malpasset test at 10 min.

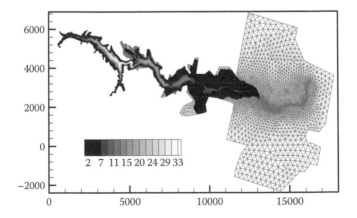

*Figure 8.41* Computed water depth for the Malpasset test at 20 min.

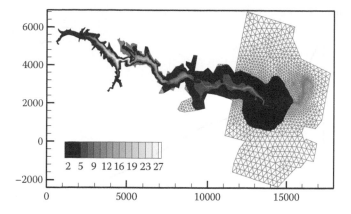

***Figure 8.42*** Computed water depth for the Malpasset test at 30 min.

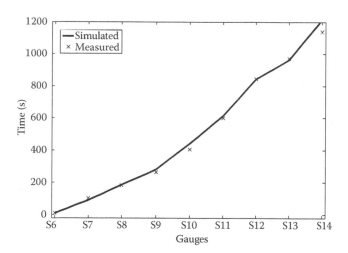

***Figure 8.43*** Computed and measured wave front arrival times for the Malpasset test.

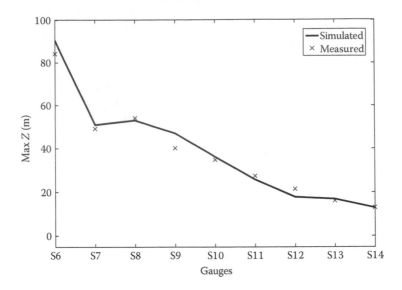

**Figure 8.44** Computed and measured maximum water levels at gauges for the Malpasset test.

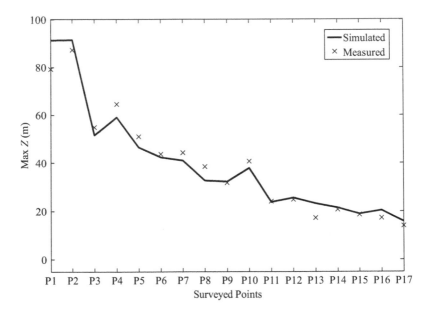

**Figure 8.45** Computed and measured maximum water levels at surveyed points for the Malpasset test.

# References

Frazão, S. S., and Testa, G. (1999). The Toce River test case: Numerical results analysis. *Proceedings of the 3rd CADAM Workshop*, Milan, Italy.

Goutal, N. (1999). The Malpasset dam failure—An overview and test case definition. Proceedings of the 4th CADAM Meeting, Zaragoza, Spain.

Hiver, J. M. (2000). Adverse-slope and slope (bump). In *Concerted Action on Dam Break Modelling: Objectives, Project Report, Test Cases, Meeting Proceedings*, edited by S. S. Frazão, M. Morris, and Y. Zech, Université catholique de Louvain, Louvain-la-Neuve, Belgium.

Sanders, B. F., and Begnudelli, L. (2010). BreZo: A hydrodynamic flood simulation algorithm. http://sanders.eng.uci.edu/brezo.html.

Thacker, W. C. (1981). Some exact solutions to the nonlinear shallow water equations. *Journal of Fluid Mechanics*, 107, 499–508.

Ying, X., Jorgeson, J., and Wang, S. S. Y. (2009). Modeling dam-break flows using finite volume method on unstructured grid. *Engineering Applications of Computational Fluid Mechanics*, 3(2), 184–194.

# chapter nine

# Pollutant transport

In this chapter, one- and two-dimensional shallow water flows with pollutant transport are considered. The governing equations and their characteristics are discussed. The details of the discontinuous Galerkin (DG) formulation are presented. Finally, several tests are conducted to show the applicability of the numerical model.

## 9.1 Pollutant transport in 1D

In this section, a numerical model for shallow water flows with pollutant transport in a rectangular channel is presented. The governing equations, DG formulation, numerical flux, initial and boundary conditions, and point source treatment are outlined in the following sections. The slope limiting scheme as described for the one-dimensional shallow water flow equations is used here and hence not repeated in this chapter.

### 9.1.1 Governing equations

For one-dimensional shallow water flows in a rectangular channel, the continuity and momentum equations are given by Equations (9.1) and (9.2), respectively. In these equations, $h$ is the water depth, $u$ is the velocity, $q = uh$ is the discharge per unit width, $S_o$ is the bed slope, $S_f$ is the friction slope, and $g$ is the gravitational acceleration. The transport equation for pollutant is given by Equation (9.3), where $T$ is the depth-averaged pollutant concentration, and $S_C$ is the depth-averaged pollutant source or sink. The transport equation can be coupled with the continuity equation and momentum equation through velocity and depth information.

$$\frac{\partial h}{\partial t} + \frac{\partial uh}{\partial x} = \frac{\partial h}{\partial t} + \frac{\partial q}{\partial x} = 0 \tag{9.1}$$

$$\frac{\partial uh}{\partial t} + \frac{\partial (hu^2 + gh^2/2)}{\partial x} = \frac{\partial q}{\partial t} + \frac{\partial \left(q^2/h + gh^2/2\right)}{\partial x} = gh(S_o - S_f) \tag{9.2}$$

$$\frac{\partial hT}{\partial t} + \frac{\partial uhT}{\partial x} = S_C \tag{9.3}$$

These three governing equations—the continuity equation, momentum equation, and transport equation—can be written in conservation vector form as given by Equation (9.4). The vector of conserved variables **U**, flux vector **F**, and vector of source terms **S** are given by Equation (9.5). The Jacobian matrix of Equation (9.4) is given by Equation (9.6) and the eigenvalues are found by solving Equation (9.7). The solution results in three distinct eigenvalues that are given by Equation (9.8). The corresponding eigenvectors are given by Equation (9.9).

$$\frac{\partial \mathbf{U}}{\partial t} + \frac{\partial \mathbf{F}}{\partial x} = \mathbf{S} \tag{9.4}$$

$$\left. \mathbf{U} = \begin{bmatrix} h \\ uh \\ hT \end{bmatrix} = \begin{bmatrix} U_1 \\ U_2 \\ U_3 \end{bmatrix}, \quad \mathbf{F} = \begin{bmatrix} uh \\ hu^2 + gh^2/2 \\ uhT \end{bmatrix} = \begin{bmatrix} U_2 \\ U_2^2/U_1 + gU_1^2/2 \\ U_2U_3/U_1 \end{bmatrix}, \right.$$

$$\mathbf{S} = \begin{bmatrix} 0 \\ gh(S_o - S_f) \\ S_C \end{bmatrix} \tag{9.5}$$

$$\mathbf{A} = \frac{\partial \mathbf{F}}{\partial \mathbf{U}} = \begin{bmatrix} 0 & 1 & 0 \\ gU_1 - \left(\dfrac{U_2}{U_1}\right)^2 & \dfrac{2U_2}{U_1} & 0 \\ -\dfrac{U_2U_3}{U_1^2} & \dfrac{U_3}{U_1} & \dfrac{U_2}{U_1} \end{bmatrix} \tag{9.6}$$

$$\det(\mathbf{A} - \lambda \mathbf{I}) = \begin{vmatrix} -\lambda & 1 & 0 \\ gU_1 - \left(\dfrac{U_2}{U_1}\right)^2 & \dfrac{2U_2}{U_1} - \lambda & 0 \\ -\dfrac{U_2U_3}{U_1^2} & \dfrac{U_3}{U_1} & \dfrac{U_2}{U_1} - \lambda \end{vmatrix} = 0 \tag{9.7}$$

$$\left. \begin{aligned} \lambda_1 &= u - \sqrt{gh} \\ \lambda_2 &= u \\ \lambda_3 &= u + \sqrt{gh} \end{aligned} \right\} \tag{9.8}$$

$$\mathbf{K}_1 = \begin{bmatrix} 1 \\ u - \sqrt{gh} \\ T \end{bmatrix}, \quad \mathbf{K}_2 = \begin{bmatrix} 0 \\ 0 \\ 1 \end{bmatrix}, \quad \mathbf{K}_3 = \begin{bmatrix} 1 \\ u + \sqrt{gh} \\ T \end{bmatrix} \tag{9.9}$$

As the governing equations have three distinct eigenvectors, the partial differential equations system is hyperbolic. For more than one pollutant species, for example, $T_1$, $T_2$, ..., $T_m$, the vector of conserved variables $\mathbf{U}$ and flux vector $\mathbf{F}$ are given by Equation (9.10). The eigenvalues for the system of $m$ different pollutant species is given by Equation (9.11).

$$\mathbf{U} = \begin{bmatrix} h \\ uh \\ hT_1 \\ hT_2 \\ \vdots \\ hT_m \end{bmatrix} = \begin{bmatrix} U_1 \\ U_2 \\ U_3 \\ U_4 \\ \vdots \\ U_{m+2} \end{bmatrix}, \quad \mathbf{F} = \begin{bmatrix} uh \\ hu^2 + gh^2/2 \\ uhT_1 \\ uhT_2 \\ \vdots \\ uhT_m \end{bmatrix} = \begin{bmatrix} U_2 \\ U_2^2/U_1 + gU_1^2/2 \\ U_2 U_3/U_1 \\ U_2 U_4/U_1 \\ \vdots \\ U_2 U_{m+2}/U_1 \end{bmatrix} \tag{9.10}$$

$$\left. \begin{aligned} \lambda_1 &= u - \sqrt{gh} \\ \lambda_2 &= \lambda_3 = \cdots = \lambda_{m+1} = u \\ \lambda_{m+2} &= u + \sqrt{gh} \end{aligned} \right\} \tag{9.11}$$

## 9.1.2 Discontinuous Galerkin formulation

In this section, the discontinuous Galerkin formulation for the hyperbolic system, given by Equation (9.4), is illustrated. The one-dimensional domain ($x = [0, L]$) is divided into $Ne$ elements, let $0 = x_1 < x_2 < \ldots < x_{Ne+1} = L$ be a partition of the domain, and a typical element is given by $I_i = [x_s^i, x_e^i]$, $1 \le i \le Ne$. Inside an element, the unknowns are approximated by Lagrange interpolation functions as given by Equation (9.12). The application of the DG formulation results in Equation (9.13). With explicit time integration, Equation (9.13) can be written for each component as given by Equation (9.14). The solution of Equation (9.14) is obtained with appropriate numerical integration and time integration as discussed in previous chapters.

$$\left. \begin{aligned} \mathbf{U} &\simeq \hat{\mathbf{U}} = \sum \mathbf{N}_j(\mathbf{x})\mathbf{U}_j(\mathbf{x},t) \\ \mathbf{F}(\mathbf{U}) &\simeq \hat{\mathbf{F}}(\mathbf{U}) = \mathbf{F}(\hat{\mathbf{U}}) \\ \mathbf{S}(\mathbf{U}) &\simeq \hat{\mathbf{S}}(\mathbf{U}) = \mathbf{S}(\hat{\mathbf{U}}) \end{aligned} \right| \tag{9.12}$$

$$\int_{x_s^e}^{x_e^e} N_i N_j \, dx \frac{\partial \mathbf{U}_j}{\partial t} + N_i \tilde{\mathbf{F}} \Big|_{x_s^e}^{x_e^e} - \int_{x_s^e}^{x_e^e} \frac{\partial N_i}{\partial x} \hat{\mathbf{F}} \, dx = \int_{x_s^e}^{x_e^e} N_i \hat{\mathbf{S}} \, dx \qquad (9.13)$$

$$\int_{x_s^e}^{x_e^e} N_i N_j \, dx \frac{\partial U_j}{\partial t} + N_i \tilde{F} \Big|_{x_s^e}^{x_e^e} - \int_{x_s^e}^{x_e^e} \frac{\partial N_i}{\partial x} \hat{F} \, dx = \int_{x_s^e}^{x_e^e} N_i \hat{S} \, dx \qquad (9.14)$$

### 9.1.3   Numerical flux

For multiple species of pollutant, the eigenvalues are given by Equation (9.15). If the Roe flux solver is used, the wave strengths $\tilde{\alpha}_i$, wave speeds $\tilde{\lambda}_i$, and eigenvectors $\tilde{\mathbf{K}}_i$ ($i = 1, 2, \ldots, m + 2$) must be determined. On the other hand, the Harten-Lax-Van Leer (HLL) solver only needs the fastest and the slowest wave speeds. In the following, the HLL solver is adopted for its simple structure. In addition, as opposed to the Roe solver, the HLL solver does not need the entropy correction (Toro, 2009). The numerical flux approximation using the HLL solver is given by Equation (9.16). The wave speeds are approximated as given by Equation (9.17).

$$\left. \begin{aligned} \lambda_1 &= u - \sqrt{gh} \\ \lambda_2 = \lambda_3 &= \cdots = \lambda_{m+1} = u \\ \lambda_{m+2} &= u + \sqrt{gh} \end{aligned} \right\} \qquad (9.15)$$

$$\mathbf{F}^{HLL} = \begin{cases} \mathbf{F}_L & \text{if} \quad S_L \geq 0 \\ \dfrac{S_R \mathbf{F}_L - S_L \mathbf{F}_R + S_L S_R (\mathbf{U}_R - \mathbf{U}_L)}{S_R - S_L} & \text{if} \quad S_L < 0 < S_R \qquad (9.16) \\ \mathbf{F}_R & \text{if} \quad S_R \leq 0 \end{cases}$$

$$\left. \begin{aligned} S_L &= \min\left(u^- - \sqrt{gh^-}, u^+ - \sqrt{gh^+}\right) \\ S_R &= \max\left(u^- + \sqrt{gh^-}, u^+ + \sqrt{gh^+}\right) \end{aligned} \right\} \qquad (9.17)$$

### 9.1.4   Initial and boundary conditions

Implementation of initial and boundary conditions follow the method of characteristics as outlined in previous chapters (Cunge et al., 1980). As there are three characteristics, three initial conditions are required. Obviously, the initial water depth, flow rate, and concentration are needed to start the

*Table 9.1* Number of Required Boundary Conditions for the
1D Transport Test

| Flow Regime | Inflow Boundary | Outflow Boundary |
|---|---|---|
| Subcritical | 2 | 1 |
| Supercritical | 3 | 0 |
| Critical | 2 | 1 |

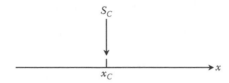

*Figure 9.1* Point source configuration.

computation. The number of required boundary conditions for the one-dimensional shallow water with pollutant transport is listed in Table 9.1.

### 9.1.5   Point source treatment

In practical problems, pollutants can be carried into rivers through pipes or sewers as point sources as shown in Figure 9.1. The point source can be viewed as a special case of concentration distribution over a very small area. The Dirac delta function satisfies the condition given by Equation (9.18). With this definition, the DG formulation corresponding to the pollutant source term is given by Equation (9.19). Note that $x_C$ represents a mesh point in the discretized domain, that is, the point source must be located at a mesh node.

$$\int_{-\infty}^{\infty} S_C(t)\delta(x - x_C)dx = S_C(t) \tag{9.18}$$

$$\int_{x_s^\ell}^{x_e^\ell} N_i \hat{S}_C \, dx = \int_{x_s^\ell}^{x_e^\ell} N_i S_C(t)\delta(x - x_C)dx = N_i(x_C)S_C \tag{9.19}$$

### 9.1.6   Numerical tests

In the first test, an idealized dam break over a flat bed with different pollutant concentrations upstream and downstream of the dam is simulated. The rectangular channel of unit width is 1000 m long with a dam located at 500 m. The initial conditions for the flow depth, velocity, and pollutant concentration are given by Equation (9.20). For the idealized dam-break problem, exact solutions for water depth and flow rate can be found (Henderson, 1966). The pollutant concentration is transported as a shock. The speed of the concentration shock, $u_C$, is given by Equation (9.21) (Audusse and Bristeau, 2003), where subscripts $L$ and $R$ denote values upstream and downstream of the dam, respectively. The values of $c_L$ and $c_R$ are given by Equation (9.22).

$$\left.\begin{aligned}
h(x \le 500 \text{ m}, t = 0) &= 10 \text{ m} \\
h(x > 500 \text{ m}, t = 0) &= 1 \text{ m} \\
u(x, t = 0) &= 0 \\
T(x \le 500 \text{ m}, t = 0) &= 0.8 \\
T(x > 500 \text{ m}, t = 0) &= 0.3
\end{aligned}\right\} \tag{9.20}$$

$$\left[\left(c_L - \frac{u_C}{2}\right)^2 - c_R^2\right]\left[\left(c_L - \frac{u_C}{2}\right)^4 - c_R^4\right] - 2u_C^2\left(c_L - \frac{u_C}{2}\right)^2 c_R^2 = 0 \tag{9.21}$$

$$\left.\begin{aligned}
c_L &= \sqrt{gh_L} \\
c_R &= \sqrt{gh_R}
\end{aligned}\right\} \tag{9.22}$$

The DG model is used to simulate the dam-break flow after instantaneous removal of the dam. The domain is discretized using 200 elements. Numerical and analytical solutions for water depth, velocity, and concentration at $t = 30$ s are compared in Figures 9.2 to 9.4. The numerical results are in good agreement with exact solutions.

In the next test, pollutant emission as a point source is simulated. The initial concentration is zero in a 1000 m long rectangular horizontal channel. At time $t = 0$ s, a source of polluted water with $S_C = 2.0$ is introduced at point $x_C = 100$ m. The initial conditions of uniform water depth and velocity are given by Equation (9.23).

$$\left.\begin{aligned}
h(x, t = 0) &= 1 \text{ m} \\
u(x, t = 0) &= 1 \text{ m/s}
\end{aligned}\right\} \tag{9.23}$$

The water depth and velocity will remain constant throughout the simulation. As mentioned in the previous section, the source point $x_C$ should be a mesh node in the simulation. The domain is discretized with

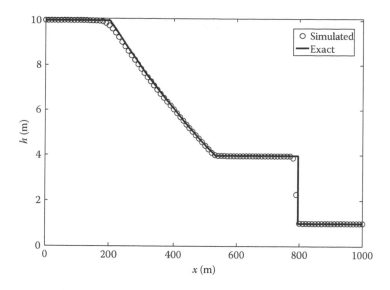

**Figure 9.2** Simulated and exact solutions of water depth after the dam break for the 1D transport test.

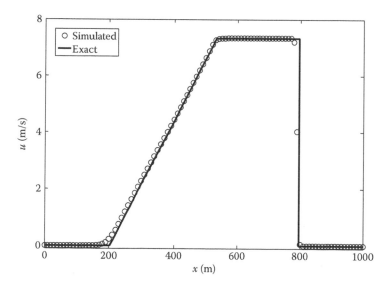

**Figure 9.3** Simulated and exact solutions of velocity after the dam break for the 1D transport test.

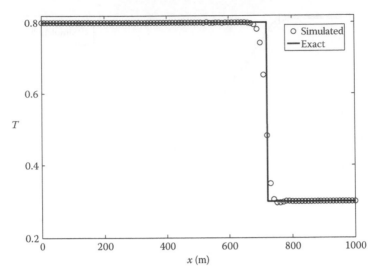

**Figure 9.4** Simulated and exact solutions of concentration after the dam break for the 1D transport test.

600 uniform size elements. The flow rate at the upstream end and water depth at the downstream end of the channel are used as boundary conditions. Two different cases are simulated in this test. In the first case, a continuous pollutant source $S_C$ is introduced, and the numerical results for concentration along the channel are shown in Figure 9.5 at $t = 300$ s

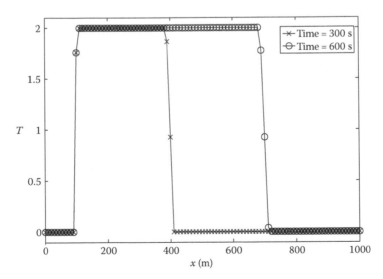

**Figure 9.5** Simulated pollutant concentration profiles with a continuous point source for the 1D transport test.

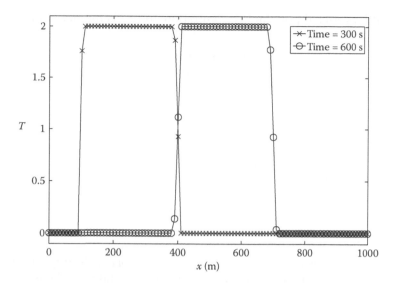

**Figure 9.6** Simulated pollutant concentration profiles after a point source deactivation for the 1D transport test.

and $t = 600$ s. In the second case, the point source is active for $0 \leq t \leq 300$ s only, afterward the released pollutant is transported by convection. The numerical results for this case are shown in Figure 9.6 at $t = 300$ s and $t = 600$ s as well. As seen from the results, the numerical results are diffusive at the shock front.

## 9.2    Pollutant transport in 2D

In this section, a numerical model for two-dimensional shallow water flows with a pollutant transport is presented. The governing equations and DG formulation are outlined. The numerical flux approximation and slope limiting schemes are exactly as described for the two-dimensional shallow water flow equations.

### 9.2.1    Governing equations

In the case of two-dimensional shallow water flows, the continuity and momentum equations are given by Equations (9.24) and (9.25), respectively. In these equations, $h$ is the water depth, $u$ and $v$ are the flow velocity along $x$ and $y$ directions, respectively, $q_x = uh$ and $q_y = vh$ are the flow rates per unit width, $S_{ox}$ and $S_{oy}$ are the bed slopes, and $S_{fx}$ and $S_{fy}$ are the friction slopes along the $x$ and $y$ directions, respectively. The transport equation of a pollutant in two-dimensional flows is given by Equation (9.26), where $T$ is the depth-averaged pollutant concentration and $S_C$ is the pollutant source or sink.

$$\frac{\partial h}{\partial t} + \frac{\partial q_x}{\partial x} + \frac{\partial q_y}{\partial y} = 0 \tag{9.24}$$

$$\left.\begin{array}{l} \dfrac{\partial q_x}{\partial t} + \dfrac{\partial\left(q_x^2/h + gh^2/2\right)}{\partial x} + \dfrac{\partial(q_x q_y/h)}{\partial y} = gh(S_{ox} - S_{fx}) \\[4mm] \dfrac{\partial q_y}{\partial t} + \dfrac{\partial(q_x q_y/h)}{\partial x} + \dfrac{\partial\left(q_y^2/h + gh^2/2\right)}{\partial y} = gh(S_{oy} - S_{fy}) \end{array}\right\} \tag{9.25}$$

$$\frac{\partial hT}{\partial t} + \frac{\partial uhT}{\partial x} + \frac{\partial vhT}{\partial y} = S_C \tag{9.26}$$

The governing equations for pollutant transport, including continuity equation, momentum equation, and transport equation, can be written in conservation form as given by Equation (9.27). The conserved variables **U**, fluxes **E** and **G**, and source terms **S** are given by Equation (9.28). The $4 \times 4$ Jacobian matrix of Equation (9.27) is given by Equation (9.29), only nonzero elements are shown. The eigenvalues and eigenvectors of the Jacobian matrix are given by Equations (9.30) and (9.31), respectively. Similar to the one-dimensional case, if there are $m$ pollutant species transported in two-dimensional shallow water, $T_1, T_2, \ldots T_m$, then the vectors of the conserved variables, fluxes, and source term are given by Equation (9.32). The Jacobian matrix will have eigenvalues as given by Equation (9.33).

$$\frac{\partial \mathbf{U}}{\partial t} + \nabla \cdot \mathbf{F} = \frac{\partial \mathbf{U}}{\partial t} + \frac{\partial \mathbf{E}}{\partial x} + \frac{\partial \mathbf{G}}{\partial y} = \mathbf{S} \tag{9.27}$$

$$\left.\begin{array}{l} \mathbf{U} = \begin{bmatrix} h \\ uh \\ vh \\ hT \end{bmatrix} = \begin{bmatrix} U_1 \\ U_2 \\ U_3 \\ U_4 \end{bmatrix}, \quad \mathbf{E} = \begin{bmatrix} uh \\ hu^2 + gh^2/2 \\ huv \\ uhT \end{bmatrix} = \begin{bmatrix} U_2 \\ U_2^2/U_1 + gU_1^2/2 \\ U_2 U_3/U_1 \\ U_2 U_4/U_1 \end{bmatrix} \\[18mm] \mathbf{G} = \begin{bmatrix} vh \\ huv \\ hv^2 + gh^2/2 \\ vhT \end{bmatrix} = \begin{bmatrix} U_3 \\ U_2 U_3/U_1 \\ U_3^2/U_1 + gU_1^2/2 \\ U_3 U_4/U_1 \end{bmatrix}, \quad \mathbf{S} = \begin{bmatrix} 0 \\ gh(S_{ox} - S_{fx}) \\ gh(S_{oy} - S_{fy}) \\ S_c \end{bmatrix} \end{array}\right\} \tag{9.28}$$

$$\left. \begin{array}{l} \mathbf{A} = \dfrac{\partial \mathbf{F} \cdot \mathbf{n}}{\partial \mathbf{U}}; A_{12} = n_x, A_{13} = n_y, A_{41} = -\dfrac{U_2 U_4 n_x}{U_1^2} - \dfrac{U_3 U_4 n_y}{U_1^2}, \\[3mm] A_{21} = \left( gU_1 - \left(\dfrac{U_2}{U_1}\right)^2 \right) n_x - \dfrac{U_2 U_3 n_y}{U_1^2}, A_{23} = \dfrac{U_2 n_y}{U_1}, A_{32} = \dfrac{U_3 n_x}{U_1}, \\[3mm] A_{31} = \left( gU_1 - \left(\dfrac{U_3}{U_1}\right)^2 \right) n_y - \dfrac{U_2 U_3 n_x}{U_1^2}, A_{42} = \dfrac{U_4 n_x}{U_1}, A_{43} = \dfrac{U_4 n_y}{U_1}, \\[3mm] A_{22} = \dfrac{2U_2 n_x}{U_1} + \dfrac{U_3 n_y}{U_1}, A_{33} = \dfrac{U_2 n_x}{U_1} + \dfrac{2U_3 n_y}{U_1}, A_{44} = \dfrac{U_2 n_x}{U_1} + \dfrac{U_3 n_y}{U_1} \end{array} \right\} \quad (9.29)$$

$$\left. \begin{array}{l} \lambda_1 = un_x + vn_y - \sqrt{gh} \\[2mm] \lambda_2 = \lambda_3 = un_x + vn_y \\[2mm] \lambda_4 = un_x + vn_y + \sqrt{gh} \end{array} \right\} \quad (9.30)$$

$$\mathbf{K}_1 = \begin{bmatrix} 1 \\ u - \sqrt{gh}n_x \\ v - \sqrt{gh}n_y \\ C/h \end{bmatrix}, \quad \mathbf{K}_2 = \begin{bmatrix} 0 \\ n_y \\ -n_x \\ 0 \end{bmatrix}, \quad \mathbf{K}_3 = \begin{bmatrix} 0 \\ n_y \\ -n_x \\ 1 \end{bmatrix}, \quad \mathbf{K}_4 = \begin{bmatrix} 1 \\ u + \sqrt{gh}n_x \\ v + \sqrt{gh}n_y \\ C/h \end{bmatrix} \quad (9.31)$$

$$\mathbf{U} = \begin{bmatrix} h \\ uh \\ vh \\ hT_1 \\ hT_2 \\ \vdots \\ hT_m \end{bmatrix}, \quad \mathbf{E} = \begin{bmatrix} uh \\ hu^2 + gh^2/2 \\ huv \\ uhT_1 \\ uhT_2 \\ \vdots \\ uhT_m \end{bmatrix}, \quad \mathbf{G} = \begin{bmatrix} vh \\ huv \\ hv^2 + gh^2/2 \\ vhT_1 \\ vhT_2 \\ \vdots \\ vhT_m \end{bmatrix} \quad (9.32)$$

$$\left. \begin{array}{l} \lambda_1 = un_x + vn_y - \sqrt{gh} \\[2mm] \lambda_2 = \lambda_3 = \cdots = \lambda_{m+2} = un_x + vn_y \\[2mm] \lambda_{m+3} = un_x + vn_y + \sqrt{gh} \end{array} \right\} \quad (9.33)$$

## 9.2.2   Discontinuous Galerkin formulation

The two-dimensional shallow water flow equations with pollutant transport equation in the conservation form are given by Equation (9.34). The computational domain is divided into $Ne$ elements as shown by Equation (9.35). Inside an element, the unknowns are approximated as given by Equation (9.36). The DG formulation for the system of equations is given by Equation (9.37), where $\tilde{\mathbf{F}}$ is the numerical flux across the element boundaries and is solved using the HLL function.

$$\frac{\partial \mathbf{U}}{\partial t} + \nabla \cdot \mathbf{F}(\mathbf{U}) = \frac{\partial \mathbf{U}}{\partial t} + \frac{\partial \mathbf{E}(\mathbf{U})}{\partial x} + \frac{\partial \mathbf{G}(\mathbf{U})}{\partial y} = \mathbf{S}(\mathbf{U}) \tag{9.34}$$

$$\Omega \approx \hat{\Omega} = \bigcup_{e=1}^{Ne} \Omega_e \tag{9.35}$$

$$\left.\begin{aligned} \mathbf{U} \approx \hat{\mathbf{U}} = \sum \mathbf{N}_j(\mathbf{x})\mathbf{U}_j(\mathbf{x},t) \\ \mathbf{F}(\mathbf{U}) \approx \hat{\mathbf{F}} = \mathbf{F}(\hat{\mathbf{U}}) \\ \mathbf{S}(\mathbf{U}) \approx \hat{\mathbf{S}} = \mathbf{S}(\hat{\mathbf{U}}) \end{aligned}\right\} \tag{9.36}$$

$$\int_{\Omega_e} \mathbf{N}_i \mathbf{N}_j \, d\Omega \frac{\partial \mathbf{U}_j}{\partial t} + \int_{\partial \Omega_e} \mathbf{N}_i \tilde{\mathbf{F}} \, d\Gamma - \int_{\Omega_e} \nabla \mathbf{N}_i \cdot \hat{\mathbf{F}} \, d\Omega = \int_{\Omega_e} \mathbf{N}_i \hat{\mathbf{S}} \, d\Omega \tag{9.37}$$

## 9.2.3   Numerical tests

The idealized dam-break problem in Section 9.1.6 is used here as a two-dimensional test. The width of the channel is 1000 m. The initial conditions are given by Equation (9.38). The simulated results at $t = 30$ s after the dam break are presented. The computed water surface and concentration profiles are shown in Figures 9.7 and 9.8, respectively. The simulated and exact solutions for the water surface, velocity, and concentration along the centerline ($y = 500$ m) of the channel are shown in Figures 9.9 to 9.11, respectively. The numerical results are in good agreement with analytic solutions and the shock waves are well captured. There is a minor difference between the simulated and exact solution in the rarefaction wave.

$$\left.\begin{aligned} h(x \leq 500 \text{ m}, t = 0) &= 10 \text{ m} \\ h(x > 500 \text{ m}, t = 0) &= 1 \text{ m} \\ u(t = 0) = 0, \ v(t = 0) &= 0 \\ \tilde{T}((x \leq 500 \text{ m}, t = 0)) &= 0.8 \\ T(x > 500 \text{ m}, t = 0) &= 0.3 \end{aligned}\right\} \tag{9.38}$$

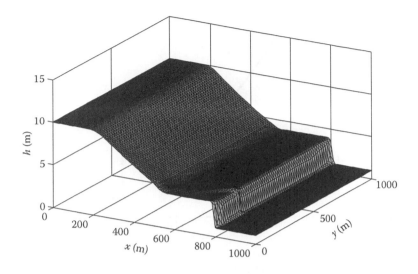

*Figure 9.7* Computed water surface profile at $t = 30$ s for the 2D transport test.

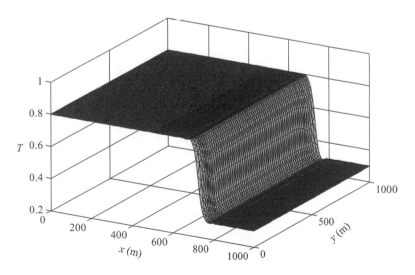

*Figure 9.8* Computed concentration profile for the 2D transport test.

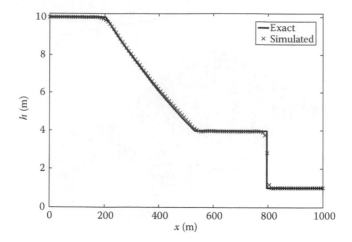

*Figure 9.9* Simulated and exact depth profiles along the centerline for the 2D transport test.

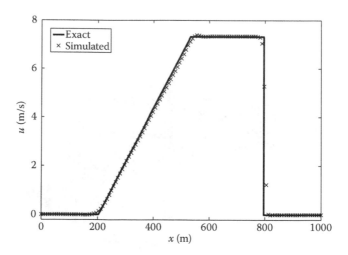

*Figure 9.10* Simulated and exact longitudinal velocity profiles along the center-line for the 2D transport test.

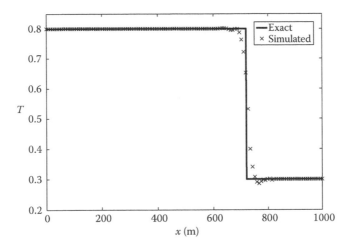

**Figure 9.11** Simulated and exact concentration profiles along the centerline for the 2D transport test.

## References

Audusse, E., and Bristeau, M. O. (2003). Transport of pollutant in shallow water, a two time steps kinetic method. *M2AN*, 37(2), 389–416.

Cunge, J. A., Holly, F. M. Jr., and Verwey, A. (1980). *Practical Aspects of Computational River Hydraulics*. Pitman, London.

Henderson, F. M. (1966). *Open Channel Flow*. McGraw-Hill, New York.

Toro, E. F. (2009). *Riemann Solvers and Numerical Methods for Fluid Dynamics*, 3rd ed. Springer-Verlag, Berlin, Heidelberg.

# chapter ten

# Concluding remarks

In this chapter, the main ideas presented in this book are summarized. Research topics and potential applications related to the work presented are discussed. Most recent developments in numerical modeling and its relevance to the DG scheme are highlighted.

## 10.1 Summary

Numerical application of discontinuous Galerkin (DG) method is introduced and presented in this book. The attention is focused on the development of total variation diminishing (TVD) Runge–Kutta discontinuous Galerkin (RKDG) method and its application to one- and two-dimensional shallow water flows.

To introduce the DG method to beginners, the DG procedure and some mathematical preliminaries are first presented in Chapter 2. Applications of the DG method to ordinary differential equations (ODEs), convection equations, and diffusion tests are presented to validate the numerical schemes.

For a broader view about the DG method, its development, theory, computation, and applications, the readers are referred to Cockburn (1997, 1999), Cockburn et al. (2000), Li (2006), Rivière (2008), and Hesthaven and Warburton (2008).

## 10.2 Current research topics

The general $P_N P_M$ approach provides a unified framework for the construction of finite volume and discontinuous Galerkin schemes (Dumbser et al., 2008; Dumbser, 2010; Gassner et al., 2011). Based on the underlying polynomial of order $N$, the time evolution and flux computation are performed with a different set of piecewise polynomials of order $M \geq N$. This approach gives a unified framework for the finite volume method ($N = 0$) and DG method ($M = N$). Numerical results for Euler equations and in the field of ideal magenetohydrodynamics (MHD) show that intermediate schemes (N $\neq 0$, $M > N$) are computationally more efficient than finite volume or DG schemes (Dumbser et al., 2008).

Slope limiters play an important role in the DG method. An ideal slope limiter should be able to preserve high-order accuracy and eliminate unphysical oscillation around shocks. In recent years, various slope limiters have been developed such as the total variation diminishing (TVD) scheme (Tu and Aliabadi, 2005; Krivodonova, 2007; Lai and Khan, 2012), and essentially nonoscillatory (ENO) and weighted essentially nonoscillatory (WENO) schemes (Dumbser et al., 2008; Zhu et al., 2008; Zhu and Qiu, 2009). The continuous development and improvement of high-order slope limiters have become essential for the discontinuous Galerkin method.

To achieve higher-order DG methods, high-order time integration methods are needed in accordance with the high-order spatial discretization. The TVD Runge–Kutta method of order up to 4 can be found in Gottlieb and Shu (1998). In the discontinuous Galerkin method, besides the explicit, nonlinear TVD Runge–Kutta schemes, the time integration can also be achieved by the Lax–Wendroff (LW) type time discretization with Cauchy–Kowalewski procedure or arbitrary high-order schemes using derivatives (ADER) type procedure. These approaches convert the time derivatives into spatial derivatives and can provide arbitrary high-order schemes as demanded. The resulting DG methods are called LWDG (Qiu, 2008) or ADER-DG (Castro and Toro, 2008) in the literature.

Computation of numerical fluxes is important in the DG scheme. A generalized Riemann problem for nonlinear hyperbolic partial differential equations with source terms is developed (Titarev and Toro, 2002; Toro, 2009). In the generalized Riemann problem, the initial conditions are approximated by arbitrary order but smooth away from the shock interface. Complete Riemann solvers, in which all the characteristic fields are accounted for in the numerical flux, remain the subject of ongoing research (Toro, 2006; Tokareva and Toro, 2010; Toro and Dumbser, 2011).

In addition to the shallow water flow problems, the discontinuous Galerkin method has been applied to a large variety of practical problems, such as Navier–Stokes equations (Bassi and Rebay, 1997; Bassi et al., 2011), Euler equations (Cockburn and Shu, 1998a; Dumbser et al., 2008), Maxwell's equations (Cockburn et al., 2004; Fezoui et al., 2005), Hamilton–Jacobi equations (Hu and Shu, 1998; Cheng and Shu, 2007), magnetohydrodynamics (Warburton and Karniadakis, 1999; Dumbser et al., 2008), elastodynamics (Chien et al., 2003; Abedi et al., 2006), multiphase flow (Sun and Wheeler, 2005; Marchandise et al., 2006), advection–reaction–diffusion system (Cockburn and Shu, 1998b; Houston et al., 2002), and heat transfer (Li, 2006).

## References

Abedi, R., Petracovici, B., and Haber, R. B. (2006). A space-time discontinuous Galerkin method for linearized elastodynamics with element-wise momentum balance. *Computer Methods in Applied Mechanics and Engineering*, 195(25–28), 3247–3273.

Bassi, F., and Rebay, S. (1997). A high-order accurate discontinuous finite element method for the numerical solution of the compressible Navier-Stokes equations. *Journal of Computational Physics*, 131(2), 267–279.

Bassi, F., Franchina, N., Ghidoni, A., and Rebay, S. (2011). Spectral p-multigrid discontinuous Galerkin solution of the Navier-Stokes equations. *International Journal for Numerical Methods in Fluids*, 67(11), 1540–1558.

Castro, C. E., and Toro, E. F. (2008). ADER DG and FV schemes for shallow water flows. In *Progress in Industrial Mathematics at ECMI 2006, Mathematics in Industry*, edited by L. L. Bonilla, M. Moscoso, G. Platero, and J. M. Vega, 12, 341–345.

Cheng, Y., and Shu, C. W. (2007). A discontinuous Galerkin finite element method for directly solving the Hamilton-Jacobi equations. *Journal of Computational Physics*, 223(1), 398–415.

Chien, C. C., Yang, C. S., and Tang, J. H. (2003). Three-dimensional transient elastodynamic analysis by a space and time-discontinuous Galerkin finite element method. *Finite Elements in Analysis and Design*, 39(7), 561–580.

Cockburn, B. (1997). An introduction to the discontinuous Galerkin method for convection-dominated problems. In *Advanced Numerical Approximation of Nonlinear Hyperbolic Equations* (Lecture Notes in Mathematics), edited by A. Quarteroni, CIME, Springer-Verlag, Berlin.

Cockburn, B. (1999). Discontinuous Galerkin methods for convection-dominated problems. In *High-Order Methods for Computational Physics* (Lecture Notes in Computational Science and Engineering, Vol. 9), edited by T. Barth and H. Deconik, 69–224, Springer-Verlag, Berlin.

Cockburn, B., Karniadakis, G. E., and Shu, C. (2000). *Discontinuous Galerkin Methods: Theory, Computation and Applications*. Springer-Verlag, Berlin Heidelberg.

Cockburn, B., Li, F., and Shu, C. W. (2004). Locally divergence-free discontinuous Galerkin methods for the Maxwell equations. *Journal of Computational Physics*, 194(2), 588–610.

Cockburn, B., and Shu, C. W. (1998a). The Runge–Kutta discontinuous Galerkin method for conservations laws V: Multidimensional systems. *Journal of Computational Physics*, 141(2), 199–224.

Cockburn, B., and Shu, C. W. (1998b). The local discontinuous Galerkin method for time-dependent convection-diffusion systems. *SIAM Journal on Numerical Analysis*, 35(6), 2440–2463.

Dumbser, M. (2010). Arbitrary high order $P_N P_M$ schemes on unstructured meshes for the compressible Navier-Stokes equations. *Computer & Physics*, 39(1), 60–76.

Dumbser, M., Balsara, D. S., Tor, E. F., and Munz, C. (2008). A unified framework for the construction of one-step finite volume and discontinuous Galerkin schemes on unstructured meshes. *Journal of Computational Physics*, 227, 8209–8253.

Fezoui, L., Lanteri, S., Lohrengel, S., and Piperno, S. (2005). Convergence and stability of a discontinuous Galerkin time-domain method for the 3D heterogeneous Maxwell equations on unstructured meshes. *ESAIM: Mathematical Modelling and Numerical Analysis*, 39, 1149–1176.

Gassner, G., Dumbser, M., Hindenlang, F., and Munz, C. (2011). Explicit one-step time discretizations for discontinuous Galerkin and finite volume schemes based on local predictors. *Journal of Computational Physics*, 203, 4232–4247.

Gottlieb, S., and Shu, C. W. (1998). Total variation diminishing Runge–Kutta schemes. *Mathematics of Computation*, 67(221), 73–85.

Hesthaven, J. S., and Warburton, T. (2008). *Nodal Discontinuous Galerkin Methods: Algorithms, Analysis, and Applications*. Springer, New York.

Houston, P., Schwab, C., and Suli, E. (2002). Discontinuous hp-finite element method for advection-diffusion-reaction problems. *SIAM Journal on Numerical Analysis*, 39(6), 2133–2163.

Hu, C., and Shu, C. W. (1998). A discontinuous Galerkin finite element method for Hamilton-Jacobi equations. ICASE Report No. 98-2.

Krivodonova, L. (2007). Limiters for high-order discontinuous Galerkin methods. *Journal of Computational Physics*, 226(1), 879–896.

Lai, W., and Khan, A. A. (2012). A discontinuous Galerkin method for two-dimensional shallow water flows. *International Journal for Numerical Methods in Fluids*, 70(8), 939–960.

Li, B. Q. (2006). *Discontinuous Finite Elements in Fluid Dynamics and Heat Transfer*. Springer-Verlag, London.

Marchandise, E., Remacle, J., and Chevaugeon, N. (2006). A quadrature-free discontinuous Galerkin method for the level set equation. *Journal of Computational Physics*, 212(1), 338–357.

Qiu, J. (2008). Development and comparison of numerical fluxes for LWDG methods. *Numerical Mathematics: Theory, Methods and Applications*, 1(4), 1–32.

Rivière, B. (2008). *Discontinuous Galerkin Methods for Solving Elliptic and Parabolic Equations: Theory and Implementation* (Frontiers in Applied Mathematics, Vol. 35). SIAM, Philadelphia.

Sun, S., and Wheeler, M. F. (2005). Discontinuous Galerkin methods for coupled flow and reactive transport problems. *Applied Numerical Mathematics*, 52(2–3), 273–298.

Titarev, V. A., and Toro, E. F. (2002). ADER: Arbitrary high order Godunov Approach. *Journal of Scientific Computing*, 17, 609–618.

Tokareva, S. A., and Toro, E. F. (2010). HLLC-type Riemann solver for Bear-Nunziato equations of compressible two-phase flow. *Journal of Computational Physics*, 229(10), 3573–3604.

Toro, E. F. (2006). Riemann solvers with evolved initial conditions. *International Journal for Numerical Methods in Fluids*, 52(4), 433–453.

Toro, E. F. (2009). *Riemann Solvers and Numerical Methods for Fluid Dynamics*. Springer-Verlag, Berlin, Heidelberg.

Toro, E. F., and Dumbser, M. (2011). Reformulated Osher-type Riemann solver. In *Computational Fluid Dynamics 2010: Proceedings of the Sixth International Conference on Computational Fluid Dynamics, ICCFD6*, edited by A. Kuzmin, Springer-Verlag, Berlin, Heidelberg.

Tu, S., and Aliabadi, S. (2005). A slope limiting procedure in discontinuous Galerkin finite element method for gasdynamics applications. *International Journal of Numerical Analysis and Modeling*, 2(2), 163–178.

Warburton, T. C., and Karniadakis, G. E. (1999). A discontinuous Galerkin method for the viscous MHD equations. *Journal of Computational Physics*, 152(2), 608–641.

Zhu, J., and Qiu, J. (2009). Hermite WENO schemes and their applications as limiters for Runge–Kutta discontinuous Galerkin method, III: Unstructured meshes. *Journal of Scientific Computing*, 39(2), 293–321.

Zhu, J., Qiu, J., Shu, C., and Dumbser, M. (2008). Runge–Kutta discontinuous Galerkin method using WENO limiter II: Unstructured meshes. *Journal of Computational Physics*, 227(9), 4330–4353.

# Index